THE WRATH OF
CAPITAL

New Directions in Critical Theory

NEW DIRECTIONS IN CRITICAL THEORY
Amy Allen, General Editor

New Directions in Critical Theory presents outstanding classic and contemporary texts in the tradition of critical social theory, broadly construed. The series aims to renew and advance the program of critical social theory, with a particular focus on theorizing contemporary struggles around gender, race, sexuality, class, and globalization and their complex interconnections.

THE WRATH OF
CAPITAL

NEOLIBERALISM AND
CLIMATE CHANGE POLITICS

Adrian Parr

Columbia University Press   New York

Columbia University Press
Publishers Since 1893
New York   Chichester, West Sussex
cup.columbia.edu

Library of Congress Cataloging-in-Publication Data
Parr, Adrian.
The wrath of capital : neoliberalism and climate change politics /
Adrian Parr.
p. cm. —(New directions in critical theory)
Includes bibliographical references and index.
ISBN 978-0-231-15828-2 (cloth : alk. paper) —
ISBN 978-0-231-53094-1 (ebook)
1. Climatic changes—Economic aspects.
2. Climatic changes—Political aspects.
3. Neoliberalism   I. Title.

QC903.P375   2013
363.738'74—dc23   2012012474

*To Deborah Zaretsky and
in memory of Allen Zaretsky*

Don't be humble; you are not that great.
—Golda Meier (1898–1978)

CONTENTS

ACKNOWLEDGMENTS

I especially thank Kenneth Surin, Jana Braziel, Annulla Linders, Marti Kheel, Thomas Heyd, and Eli Zaretsky for their thoughtful suggestions and comments. I am extremely grateful to Jack Elliott and the Hillier Endowment at Cornell University; Jill Bennett, Felicity Fenner, and Dennis del Favero at the University of New South Wales; and Joanne de Vries from the Fresh Outlook Foundation for providing me with opportunities to share earlier drafts of the manuscript. Sheryl Cohen kept me on track throughout every phase of this project, and I cannot thank her enough for her unwavering support.

Stimulating conversation with Nancy Fraser, Raj Patel, Michael Hardt, Debal Deb, Charles Waldheim, Stephen Seidel, Daniel Greenberg, Durganand Balsavar, Bryony Schwan, Noel Sturgeon, Greta Gaard, Mike Parr, Antoni Fokers, all the people I met in Dharavi, and the rural farmers outside Hyderabad, India, informed the arguments I develop throughout the book. I am very appreciative of my colleague Nnamdi Elleh for his continuing encouragement. My research assistants Erin Gulley and Katherine Setser worked hard investigating scientific evidence and reports; Katy Johnson's eye for detail was an enormous help in the latter stages of the editing process; and Emily Schweppe saved the day on more than one occasion.

The University of Cincinnati Faculty Development Council, the University of Cincinnati Research Council, and the Taft Center for Research on the Humanities provided me with much needed funding to complete research for this book. I also thank Wendy Lochner and Amy Allen for their enthusiasm for this project as well as Christine Mortlock and copy-editor Annie Barva for their assistance at various stages of the publication process.

Last, I am very lucky to have Michael, Lucien, Shoshana, and Yehuda in my life; without them, this book would never have come to be.

ABBREVIATIONS

BP	British Petroleum
CCC	Commercial Club of Chicago
CCX	Chicago Climate Exchange
CDM	Clean Development Mechanism
CO_2	carbon dioxide
ERU	emission-reduction unit
EU ETS	European Union Emissions Trading Scheme
FACE	Forests Absorbing Carbon Dioxide Emissions
FAO	Food and Agriculture Organization
FDA	U.S. Food and Drug Administration
GDP	gross domestic product
GHG	greenhouse gas
HFC-23	trifluoromethane
HOPE	Housing Opportunities for People Everywhere
IPCC	Intergovernmental Panel on Climate Change
JI	Joint Implementation
LEED	Leadership in Energy and Environmental Design
NAFTA	North American Free Trade Agreement
NGO	nongovernmental organization
ppm	parts per million
STRASA	Stress Tolerant Rice for Africa and South Asia
UN	United Nations
USGBC	U.S. Green Building Council
UWA	Uganda Wildlife Authority
WHO	World Health Organization

THE WRATH OF
CAPITAL

INTRODUCTION

BUSINESS AS USUAL

One can only speak of what is in front of him, and that is
simply a mess.

—Samuel Beckett[1]

A legend dating back to sixteenth-century Prague tells of a violent conflict between a man and his creation. The story begins with the rabbi
of Prague, Judah Loew (1525–1609), who in a dream is instructed
to create a golem to protect his people from blood libel.[2] Heading down to
the muddy banks of the Moldau River, the rabbi and his two assistants create a humanoid form out of earth, fire, water, and air, bringing the clay figure to life. The golem would consign the blood libel to the trash of history,
and from there civilization could continue forward in peace and harmony.
But the titanic muscular clay creature becomes increasingly independent,
unmanageable, and injurious, grasping the city tightly in its fists, terror-
izing everyone.

The rabbi tries to cooperate with the creature in hopes that by demon-
strating his love for the golem, it will become more self-reflexive, capable
of distinguishing between care and indiscriminate destruction. Unfortunately, with each day that passes the golem becomes increasingly uncompromising and harmful, posing a seemingly intractable problem for the
rabbi. Although it shields the rabbi's people from attack, it annihilates
everything else in its path. Exhausted by the situation and the growing

The Golem

debris, the rabbi is compelled to act. As painful as it is, he returns back to the earth the creature he had authored into existence.

The fable provides an intriguing perspective on freedom and autonomy. The golem has no freedom: it is the rabbi who brings it to life and sentences it to death. Yet by returning the creature to earth, the rabbi holds the golem accountable for the destruction it wrought despite not being free. This is the basic premise of this book. We are not free, yet we are autonomous. We are constrained by the historical circumstances into which we are born, along with the institutions and structures that contain us. Nonetheless, each and every one of us also participates in and thereby confirms the legitimacy of those selfsame institutions and structures that dominate us, along with the violence they sustain.[3] In this way, we are both the rabbi creator and the creature creation. Insofar as we are socially constituted, we are constrained by the historical and institutional forces that construct us. As political agents, we realize our autonomy as we interrupt and contest the historical and institutional conditions that regulate and organize the frames of reference through which we think and act. This structure of rupture and continuity is the modern narrative par excellence.

Fredric Jameson neatly summarizes the narrative condition of modernity as the dialectic between the modality of rupture that inaugurates a new period and the definition of that new period in turn by continuity.[4] The ironical outcome, as I describe it in the pages that follow, is that despite the narrative category driving change in the modern world, everything continues to stay the same—perhaps because what this narrative produces is a virulent strain of amnesia. Every change or historical rupture contains within it the dialectical narrative structure of modernity such that the New and the period it launches into existence are mere ritual. What persists is the condition of violence embedded in neoliberal capitalism as it robs each and every one of us (other species and ecosystems included) of a future.

The narrative of modernity and the optimistic feeling of newness it generates are merely a distraction. Distractions such as decarbonizing the free-market economy, buying carbon offsets, handing out contraceptives to poor women in developing countries, drinking tap water in place of bottled water, changing personal eating habits, installing green roofs on city hall, and expressing moral outrage at British Petroleum (BP) for the oil spill in the Gulf of Mexico, although well meaning, are merely symptomatic of the uselessness of free-market "solutions" to environmental change. Indeed, such widespread distraction leads to denial.

With the proclamation of the twenty-first century to be the era of climate change, the Trojan horse of neoliberal restructuring entered the political arena of climate change talks and policy, and a more virulent strain of capital accumulation began. For this reason, delegates from the African nations, with the support of the Group of 77 (developing countries), walked out of the 2009 United Nations (UN) climate talks in Copenhagen, accusing rich countries of dragging their heels on reducing greenhouse gas (GHG) emissions and destroying the mechanism through which this reduction can be achieved—the Kyoto Protocol. In the absence of an internationally binding agreement on emissions reductions, all individual actions taken to reduce emissions—a flat global carbon tax, recycling, hybrid cars, carbon offsets, a few solar panels here and there, and so on—are mere theatrics.

In this book, I argue that underpinning the massive environmental changes happening around the world, of which climate change is an important factor, is an unchanging socioeconomic condition (neoliberal capitalism), and the magnitude of this situation is that of a political crisis. So, at the risk of extending my literary license too far, it is fair to say that the human race is currently in the middle of an earth-shattering historical moment. Glaciers in the Himalayas, Andes, Rockies, and Alps are receding. The social impact of environmental change is now acute, with the International Organization for Migration predicting there will be approximately two hundred million environmental refugees by 2050, with estimates expecting as many as up to one billion.[5] We are poised between needing to radically transform how we live and becoming extinct.

Modern (postindustrial) society inaugurated what geologists refer to as the "Anthropocene age," when human activities began to drive environmental change, replacing the Holocene, which for the previous ten thousand years was the era when the earth regulated the environment.[6] Since then people have been pumping GHGs into the atmosphere at a faster rate than the earth can reabsorb them. If we remain on our current course of global GHG emissions, the earth's average climate will rise 3°C by the end of the twenty-first century (with a 2 to 4.5° probable range of uncertainty). The warmer the world gets, the less effectively the earth's biological systems can absorb carbon. The more the earth's climate heats up, the more carbon dioxide (CO_2) plants and soils will release; this feedback loop will further increase climate heating. When carbon feedback is factored into the climate equation, climate models predict that the rise in average climate temperature will be 6°C by 2100 (with a 4 to 8°C probable range of uncertainty).[7] For this reason, even if emissions were reduced from now on by approximately 3 percent annually, there is only a fifty–fifty chance that we can stay within the 2°C benchmark set by the UN Intergovernmental Panel on Climate Change (IPCC) in 2007. However, given that in 2010 the world's annual growth rate of atmospheric carbon was the largest in a decade, bringing the world's CO_2 concentrations to 389.6 parts per million (ppm) and pushing concentrations to 39 percent higher than what they were in 1750 at the beginning of the Industrial Revolution (approximately 278 ppm), and that there is no sign of growth slowing, then even the fifty–fifty window of opportunity not to exceed 2°C warming is quickly closing. If we continue at the current rate of GHG emissions growth, we will be on course for a devastating scenario.[8] We need to change course *now*.[9]

The Problem

Climate change poses several environmental problems, many of which now have a clear focus. The scientific problem: How can the high amounts of CO_2 in the atmosphere causing the earth's climate to change be lowered to 350 ppm? The economic problem: How can the economy be decarbonized while addressing global economic disparities? The social problem: How can human societies change their climate-altering behaviors and adapt to changes in climate?[10] The cultural problem: How can commodity culture be reigned in? The problem policymakers face: What regulations can be introduced to inhibit environmental degradation, promote GHG reductions, and assist the people, species, and ecosystems most vulnerable to environmental change? The political problem is less clear, however, perhaps because of its philosophical implications.

Political philosophy examines *how* these questions are dealt with and the assumptions upon which they are premised. It studies the myriad ways in which individuals, corporations, the world's leaders, nongovernmental organizations (NGOs), and communities respond to climate change and the larger issue of environmental change characteristic of the Anthropocene age. More important, political philosophy considers how these responses reinforce social and economic structures of power. In light of this consideration, how do we make the dramatic and necessary changes needed to adapt equitably to environmental change without the economically powerful claiming ownership over the collective impetus and goals that this historical juncture presents?

By drawing attention to the political problem of equality in the context of environmental change, I need to stress that I am not a market Luddite; rather, I am critical of the neoliberal paradigm of economic activity that advances deregulation, competition, individualism, and privatization, all the while rolling back on social services and producing widespread inequities and uneven patterns of development and social prosperity. I am also not intending to make negotiable the "non-negotiable planetary preconditions that humanity needs to respect in order to avoid the risk of deleterious or even catastrophic environmental change at continental to global scales."[11] Indeed, my argument is that by focusing too much on free-market solutions to the detriment of the world's most vulnerable (the poor, other species, ecosystems, and future generations), we make these preconditions negotiable: the free market is left to negotiate our future for us.

The contradiction of capitalism is that it is an uncompromising structure of negotiation. It ruthlessly absorbs sociohistorical limits and the challenges these limits pose to capital, placing them in the service of further capital accumulation. Neoliberalism is an exclusive system premised upon the logic of property rights and the expansion of these rights, all the while maintaining that the free market is self-regulating, sufficiently and efficiently working to establish individual and collective well-being. In reality, however, socioeconomic disparities have become more acute the world over, and the world's "common wealth," as David Bollier and later Michael Hardt and Antonio Negri note, has been increasingly privatized.[12] In 2010, the financial wealth of the world's high-net-worth individuals (with investable assets of $1 to $50 million or more [all money amounts are in U.S. dollars]) surpassed the 2007 pre–financial crisis peak, growing 9.7 percent and reaching $42.7 trillion. Also in 2010 the global population of high-net-worth individuals grew 8.3 percent to 10.9 million.[13] In 2010, the global population was 6.9 billion, of whom there were 1,000 billionaires; 80,000 ultra-high-net-worth individuals with average wealth exceeding $50 million; 3 billion with an average wealth of $10,000, of which 1.1 billion owned less than $1,000; and 2.5 billion who were reportedly "unbanked" (without a bank account and thus living on the margins of the formal financial system).[14] In a world where financial advantage brings with it political benefits, these figures attest to the weak position the majority of the world occupies in the arena of environmental and climate change politics.

Neoliberal capitalism ameliorates the threat posed by environmental change by taking control of the collective call it issues forth, splintering the collective into a disparate and confusing array of individual choices competing with one another over how best to solve the crisis. Through this process of competition, the collective nature of the crisis is restructured and privatized, then put to work for the production and circulation of capital as the average wealth of the world's high-net-worth individuals grows at the expense of the majority of the world living in abject poverty. Advocating that the free market can solve debilitating environmental changes and the climate crisis is not a political response to these problems; it is merely a political ghost emptied of its collective aspirations.

In the following pages, I mine the political and pragmatic implications of this dance of death between neoliberal capitalism and environmental change. I prefer to use the term *environmental change* rather than *climate change* except when directly dealing with the issue of CO_2 buildup in the atmosphere. When I use the term *climate change*, I am specifically referring to the long-term warming of the earth as a result of GHGs entering the atmosphere because of human activities. The "changes" that the term *environmental change* refers to are both the changes that are the result of human activities' thickening the earth's CO_2 blanket and the broader environmental changes wrought by modernity and the free market, such as the privatization of the commons, landfills, freshwater scarcity, floods, desertification, landslides, coastal and soil erosion, drought, crop failures, extreme storm activity, land degradation and conversion for agriculture and livestock farming, urban heat-island effect, polluted waterways, ocean acidification, and many other problems on a growing list.

We might not be able to see, hear, smell, taste, or touch the climate as it changes. Most of us—those who do not have the sophisticated empirical skills and knowledge of a scientist—cannot directly witness changes in the earth's climate. But all of us are a part of history, and as such we bear witness to the changes in our environment every day of our lives.

I notice that the summers are becoming unbearably hot and that the coldest days in winter are brutally cold. I am surprised when leaves fall later in the year or when my tomato has no taste or if my local fish restaurant announces that "cod is temporarily unavailable" because fish stocks are low. In the Colorado Rockies, I am alarmed at all the dead trees wiped out by a particularly virulent outbreak of mountain pine beetle. I am distressed when my city of birth, Sydney, looks like planet Mars after being covered in a blanket of orange dust from the outback. I am appalled by documentary footage showing factory-farm animals not far from where I live in Ohio being tortured and the toxic sludge that large-scale livestock farming reduces the land to, so much so that a piece of steak has never looked or tasted the same again, prompting me to stop eating meat altogether. As I walk for miles in rural India in search of potable water, I realize how lucky I am that I usually have the luxury of just turning on the tap; I cannot say the same for the barefoot, rag-clad children who beg me to buy a bottle of water for them as well. I am dismayed when Ohio governor John Kasich signs into law (July 1, 2011) a new measure that will open up our state parks to oil and gas drilling. The environment in which I live at this moment in time structures how I understand and encounter others and the world in which I live; all of us (other people, species, ecosystems, future generations, and I) are part of this broader collective life and time.

Environmental change exposes problems inherent to the modern political order and presents that order with a crisis. Although this book is very much about the failure of politics to produce equitable political options in response to environmental change, it is also an attempt to break through the dominant political edifice to get to the structures and conditions that constrain a viable alternative from appearing. Habitual thinking and praxis have to be replaced by a more utopian imagination—one that injects disobedience into the institutionalized political order.

The philosophical concerns that drive my analysis are the failure of imagination, the poverty of politics, the nature of change, and the meaning of life in the absence of a future. I suggest that this political crisis concerns the distant relationship between utopian imagination (ideal futures) and social unrest (real presents and pasts) as well as, more significant, the new collective arrangements that the utopian imagination and social unrest create when brought into proximity with each other.

All in all, my point is that it matters *who* claims ownership of the discourse and politics surrounding environmental and climatic change and *how* they do so. One significant political lesson we can take away from the failure of Kyoto and of the various international climate change talks over the past few decades is that if the economically powerful are allowed to continue monopolizing the meaning of environmental change, then the disagreement and disobedience that collective conditions and aspirations present lose their relevance.

1

CLIMATE CAPITALISM

t is now widely accepted, apart from by a few conservative fundamental-
ists and conspiracy theorists such as Lord Monckton, that the average
global climate is warming and that one of the primary causes for this
situation is human activities, which are producing more GHGs than the
earth's carbon sinks can absorb. Industrialization and a rampant culture of
consumption have resulted in the warming of the earth's atmosphere and
oceans.[1] And it is no longer just the scientists who are worried; the general
public has started to sit up and take note of climate warming. So why all
this concern over a few changes in degrees?

A few snippets clearly bring the climate situation into focus. If the
Greenland or West Antarctic ice sheets collapse, the sea level will rise several meters, causing entire islands and coastline communities to disappear.
The IPCC has estimated that the global average sea-level rise might be
anywhere from 0.18 to 0.59 meters.[2] In 2009, at the International Scientific
Congress on Climate Change, it was reported that "the upper range of sea
level rise by 2100 could be in the range of about one meter."[3] In addition,
desertification and drought may cause agriculture yields to drop. And, as
noted in the *Stern Review*, the food crisis alone may very well lead to a rise

in conflict.[4] In addition, there will be ongoing species extinction, and the incidence of extreme weather events will increase.

In an effort to stave off the catastrophic effects climate change will have for life on earth, one proposed and rather popular solution has been to try to keep the increase in global temperatures to 2°C above late preindustrial levels. Meanwhile, small island states have been demanding this target be revised to an increase that does not exceed 1.5°C.[5] Small island states insist that GHG emissions need to be reduced by more than 85 percent below 1990 levels by 2050.[6] As the science of climate change continues to influence the spheres of law, policy, economic development, and cultural production, a new debate over how to lower global GHG emissions ethically is gathering momentum. For many, when it comes to answering the difficult question concerning who should bear the burdens associated with lowering GHG emissions and the subsequent problem of how to reconcile the conflicting ethical issues this topic raises, the principle of socioeconomic justice is often invoked: the solutions need to be fair and equitable.

The issue of socioeconomic inequality has come to animate recent discussions concerning the ethics of climate change. First, the socioeconomic disparity thesis appears in arguments used against global compliance for constraints on emissions. The position of nonglobal compliance on emissions argues that because some countries—namely, high-income ones such as the United States, the United Kingdom, and other European Union (EU) countries—are better positioned in terms of economic wealth, economic strength, and technological know-how, they will incur far less hardship if they cut back on their emissions than will low- or middle-income countries.[7] This view is called the "ability-to-pay principle," and it claims that the privileged position that high-income countries enjoy constitutes a moral responsibility to support nonglobal compliance. As the argument goes, poorer countries would incur a disproportionate hardship if they slowed their economies.

Henry Shue maintains that not reducing GHG emissions amounts to a violation of a duty to not harm future generations.[8] He supports a ratio system of measurement for how much GHGs a country ought to be allowed to emit. We ought, he says, consider a country's available resources when deciding *who* should pay *what*. The principle of equity espoused by Shue is also one that appears in the Kyoto agreement. It basically responds to the problem of responsibility by invoking a concept of redistribution. Egalitarian redistribution in the context of climate change ethics both demands that wealthier nations foot more of the bill because they have the resources to do so and appeals to the principle of historical responsibility.

Another line of reasoning that is structured by the principle of historical responsibility is the call to make polluters pay. Advocates of this approach argue that those who produce pollution are responsible for the costs incurred by the damage. As a follow-on from the "polluter pays" principle is the argument that because the poverty encountered by many middle- and low-income countries is largely a by-product of past wrongs inflicted on them by high-income countries, such as the United States and former colonial powers including the United Kingdom, which have been profit-ing from years of industrial activity, this history alleviates poor countries' moral responsibility to comply with global constraints on emissions and places the ball of responsibility squarely in the court of wealthy countries. In particular, it is strongly argued that the United States should take on more of the burdens associated with climate change.[9]

Peter Singer has suggested that the developing world turn a blind eye to the developed world's historical responsibility. He recognizes that there is a limit to the amount of GHGs that the planet can absorb and suggests that one way to solve the problem of GHG buildup is to reach a consen-sus over what subsistence levels of emissions are. From this consensus, we would be able to allocate the amount of carbon each individual can safely emit. Using 2002 as his benchmark, he proposes that every person can emit one metric ton of carbon per year.[10] However, Singer's dismissal of histori-cal contingency is grounded in his analytic approach to problem solving, a way of reasoning that frames problems in isolation from the vicissitudes of economic, social, political, and cultural life. His position is basically con-tradictory. Normative arguments of justice aside, it is inconsistent to argue in favor of erasing historical responsibility in order to achieve historical responsibility—the responsibility for future lives. This is like having your cake and eating it too.

Another version of the "polluter pays" principle is the position advanced by Robin Attfield and George Monbiot. They maintain that we need to intro-duce a system of carbon rationing calculated by using population figures.[11] This system would enable a fairer distribution of the economic burdens incurred. For instance, the United States and the United Kingdom would have to reduce their carbon emissions dramatically, but lower-income coun-tries would not. According to Monbiot, the target for 2030 would entail that high-income countries cut their GHG emissions by 90 percent.[12]

Polluter Pays

Similar to Singer's stabilization thesis and also in support of those who advocate emissions-trading schemes, the Global Commons Institute advances a contraction-and-convergence approach to the problem of global climate change. It, too, leans upon a historically constituted principle of social equity insofar as it aspires to narrow the gap between the wealthy and the poor. The theory aims to produce equal per capita emissions and favors emissions trading to get there. First, a figure for a safe level of global GHG emissions needs to be set. Second, these emissions would converge to form the basis of per capita quotas. The principle of socioeconomic distribution would come into effect in that wealthy countries would need to contract their emissions more than poorer countries. In addition, poor countries might initially be allowed to increase their emissions. From here, total global emissions would begin to contract.[13]

These arguments might not be perfect, but they do offer up a road map to cutting carbon emissions across the globe. Why, then, cannot the leaders of the world reach a consensus? The question is almost a naive one to ask because the answer is so obvious. Cutting carbon emissions will hurt the economy—that is, unless the economy can be tweaked in such a way that it capitalizes from climate change. Interestingly enough, the latter argument is gaining traction in the form of "climate capitalism." My use of the term *climate capitalism* is intended to be tongue in cheek. I am fully aware of how it is gaining popularity among scholars and policymakers who hope to put the mechanisms of capitalism to work in the service of decarbonizing the economy,[14] but I disagree with them. As I say often in this book, capitalism appropriates limits to capital by placing them in the service of capital; in the process, it obscures the inequities, socioeconomic distortions, and violence that these limits expose, thereby continuing the cycle of endless economic growth that is achieved at the expense of more vulnerable entities and groups.

On December 1, 2008, IPCC chairman Rajendra Pachauri explained the connection between environmental well-being and human well-being, outlining the ways in which the current global economic system is implicated in furthering climate change. Preempting his critics, who exclaim that serious reductions would cripple the global economy, Pachauri has argued that the costs associated with reducing GHG emissions will be less than 3 percent of global gross domestic product (GDP) in 2030. However, this estimate does not factor into the equation savings incurred as a result of stabilizing the earth's global temperature. If we consider these savings, we may be surprised to find that economic output and welfare increase overall.[15] Pachauri argues that we can slow or even perhaps reverse climate change through green technologies and by the transformation of dirty industries into green ones. In turn, these changes can provide the basis for new wealth production.

Other leading figures from the sustainability movement also view climate change as a terrific opportunity to open up the market to new sources of growth by using the climate crisis as another instrument through which neoliberal economic policies can be bolstered. William McDonough and Michael Braungart offer a new design framework that the corporate sector can adopt to produce commodities that are biological nutrients and that are manufactured without the use of toxic materials.[16] They provide practical ways to achieve this goal, and the model they describe is rich with possibilities to transform the life cycle of everyday commodities. Their vision of industry in the future is one that is smart, effective, harmless, and inspiring. They provide a hopeful picture of a new green production process that no longer produces waste and instead produces nutrients for life on earth. However, the fundamental principles of neoliberal economics—privatization, competition, deregulation—remain intact.

As strong supporters of climate capitalism, L. Hunter Lovins and Boyd Cohen advocate that market mechanisms be put to use to solve the climate crisis. In support of their idea, they cite evidence from the British Carbon Disclosure Project, which in 2007 reported that the world's "major companies are increasingly focused on climate change and . . . see it as an opportunity for profit." They cover a wide range of very helpful practical options to respond to the climate crisis, such as producing clean technologies, sustainable manufacturing processes, energy-efficient buildings, a carbon-offset market that individuals can voluntarily participate in—all of which they stipulate can be promoted under the rubric of an "honest free market" that has learned the hard lesson of market humility. Much like Alan Greenspan was forced to concede during the peak of the global financial crisis that markets need some regulation, so, too, Lovins and Cohen explain that "markets make very good servants, but they're not good masters, and they're a lousy religion."[17] Despite this admission, they keep the neoliberal emphasis on the privatization and entrepreneurial self-interest that characterize capitalism and that have also produced widespread inequity the world over. As such, their notion of climate capitalism leaves unchallenged the violence embedded in capital accumulation.

If we broach the myriad problems associated with climate and environ-mental change through the lens of capitalism and market growth, aren't we coming at the problem from the wrong vantage point? If economic pros-perity is viewed in terms of geopolitical power and within the neoliberal economic paradigm of privatization, competitive markets, and deregula-tion in the way advocated by Thomas Friedman, whom Lovins and Cohen lean on extensively to help make their point, then there is really nothing new being proposed. The problem of climate change, GHG emissions, fos-sil fuel energy, pollution, deforestation, species extinction, and ecosystem degradation is being situated in a capitalist context, leaving the axiomatic of capital unchallenged and along with it the inequities that such a system produces. Climate change and environmental degradation, however, are ultimately problems of equality. All people (including future generations), other-than-human species, and ecosystems share a common future.

To argue that climate change offers opportunities for new markets and wealth production ironically reproduces the normative neoliberal value in commodity culture. For this reason, detaching inequitable socioeco-nomic arrangements from an analysis of how power dynamically operates throughout the domain of commodity culture and the consumer credit economy is misleading. Regardless of how robust the new "green free mar-ket" is, changes in climate will cause serious societal disruption, and it will be the poor (regardless of what country they live in) who will bear the greater burden in this regard.[18]

We must recognize that global heating will not affect everyone equally. It will irreversibly change environments, which will also threaten people's livelihoods. As a consequence, conflict will almost certainly increase. Some glaciers, such as those in the Andean region, are now under threat if cur-rent rates of warming persist. This situation may result in decreased river flows that will affect 500 million people in South Asia and around 250 mil-lion in China.[19] As I argued in *Hijacking Sustainability*,[20] all in all, the poor of the world are especially vulnerable in the face of these changes. Global heating is not indiscriminate; its effects will more severely impact already marginalized and exploited groups. The point in all this is that the aver-age change in global climate concerns the material conditions of life, which in turn cannot be understood without consideration of how the political economy shapes climate change politics; in other words, the production of green technologies, green energy, green jobs, and social arrangements are intertwined phenomena.

What is missing in the analyses outlined thus far is a return to Marx's *Capital* and the simple idea that the so-called opportunities in question are not isolated empirical incidents; their very existence has come into being by virtue of the political economy. How the value of these new opportunities in the context of climate change is assessed is not just in terms of their having a use-value (through the redistribution of wealth), but also in terms of their surplus-value (the production of profit). The surplus-value of the opportunities in question cannot be defined in isolation from other opportunities (green jobs, a growing consumer market in natural and green products, and new green technologies). The value of a surplus emerges through a system of exchange, and it is the comparative aspect of this Marxist idea of exchange that demands a more critical evaluation of how the political economy in the era of climate change operates.[21]

In 2004, the twenty-five countries with the largest GHG emissions accounted for 83 percent of global emissions, the United States producing 20.6 percent of those emissions. At that time, the United States, along with the twenty-five-member state body of the EU, also enjoyed generating 21.9 percent of global GDP. Although in 2005 China's emissions were 2 percent lower than those of the United States, by 2006 China surpassed the United States by 8 percent, and in 2007 two-thirds of the world's 3.1 percent increase in CO_2 was a result of China's emissions.[22] China's increasing GHG emissions obviously pose a serious problem for the goal of reducing global GHG emissions.

U.S. carbon emissions are largely the result of the energy (combustion of fossil fuels) required to power the world's largest economy, and the production processes attributed to a great deal of these energy-related CO_2 emissions have been outsourced to China, so there is good reason to suggest that the figures describing China's increase in emissions unfairly point the finger of blame at China and divert attention away from the United States (as well as from other countries such as Australia, Japan, and the wealthy Arab nations). In other words, the increase in China's emissions are most likely the result of high-income countries such as the United States outsourcing their own dirty manufacturing to China. In support of this claim, it is useful to compare the two countries' per capita CO_2 emissions. In 2007, the United States ranked highest in emissions in the world, emitting 19.4 metric tons of CO_2 per capita, whereas China emitted 5.1 metric tons of CO_2 per capita despite the fact that China's output includes the GHGs emitted as a result of wealthier nations outsourcing their production to China. The thesis of carbon rationing according to population size attempts to address this discrepancy. It alerts us to the importance of incorporating an analysis of political economy into the discourse of climate change. It prompts us to probe a little more the connection between the division of money, labor, and value driving the global consumer economy.

In an effort to save costs, more and more U.S. corporations are offshoring and outsourcing manufacturing and services to middle-income countries such as India and China where labor is cheap. For example, in 2005 manufacturing employment in the United States was 19 percent lower than in 1998 regardless of a 10 percent increase in manufacturing output.[23] And although outsourcing and offshoring results in the dispersal of productivity and GHG emissions across the globe, there is a concomitant centralization of capital accumulation occurring in the epicenters of global corporate power.

A rising powerful transnational industry is emerging around the research and development of new technologies that are designed to offset and even save the human race from global heating. Solar thermal power, electricity generated from wind, and, more recently, advances in geoengineering (one such initiative reflects solar heat away from the earth) are just a few examples of new green products entering the market, some of which I critically evaluate in chapter 2 on the voluntary carbon-offset economy. I am modestly optimistic about these technologies and hope we will continue to put the power of human imagination and problem solving to work in ways that might help mitigate climate change and even make life on a heating planet worth bearing. That said, as we work to build a green free-market economy, where is the incentive to question the economic framework that is used as the basis for green technological advancement? It is important to ask the following questions: Who will "own" the patents on new green technologies? Who will have access to them, and how much will access cost?

A few preliminary questions should be raised around the problem of the political economy in the context of solar power technologies. If solar energy is part of our common wealth, are the technologies used to access this resource also part of the global commons? Or are these technologies part of the free-market economy? If they are, then solar power is just another commodity to be bought and sold on the free market. Solar technologies are more effective in certain contexts, such as the United States, southern Europe, and Australia. Given that solar efficiency decreases the farther you go from the equator, where does this leave countries such as Russia and Norway? How do you offset the cost of solar panel installation? Does this mean poor countries with an abundance of solar power will attract solar power investment? And what are the terms and conditions of such investment? There is the great potential for solar power to be generated in the developing countries of the African continent, but such a project would require enormous financial and political investment. Who will bear the burden of this investment, and how are the benefits to be calculated and distributed? In addition, how can Africans be assured that such investment in their countries will not end up as a distorted form of colonization? In chapter 8, I raise concerns over the geopolitics of U.S. HIV/AIDS programs for Africans, looking at the ways in which such humanitarianism is implicated in the geopolitics of oil capitalism. These questions invite consideration of the political economy and its neoliberal manifestation in particular.

Neoliberalism has its roots in the political philosophy of Adam Smith, who argued in favor of curbing government restrictions and removing the barriers to economic growth. As an economic system, liberalism really kicked in throughout the 1900s, suffering only a brief setback during the Great Depression of the 1930s. With the laissez-faire policies of U.K. prime minister Margaret Thatcher (1979–1990) and the fortieth U.S. president Ronald Reagan (1981–1989), a more virulent strain of liberalism emerged during the 1980s, hence the *neo* in *neoliberalism*. By and large, neoliberalism has bastardized the fundamentals of liberalism—namely, freedom, rights, and individual choice. In the name of celebrating individual responsibility and choice, neoliberal policies have resulted in cutbacks on government spending, mass privatization, trickle-down economics, deregulation, open competition, and the gradual deterioration of the commons.[24]

A green free market cannot be presented as the solution that will automatically—pardon the analogy here—kill two birds with one stone (new opportunities for market growth and cuts in emissions). The green free market favors the current system of privatization at the expense of exploring new economic alternatives; for this reason, it is mere cronyism. It panders to neoliberal forces by commercializing global heating. In so doing, it reinforces the structural distortions of economic neoliberalism. There is no direct correlation between global green economic output and socioeconomic equity, which will happen only if there are mechanisms in place that can recognize that a green economy has to be fine-tuned in a way that allows it to become a transformative force that can prompt the development of economic opportunities for the poor, be a means for wealth redistribution, and shift the dominant cultural value away from privileging competition, private-property ownership, and wealth accumulation.

An economy is not just a mode of production; it is also a productive force. It has the potential to change the current material conditions of life that constitute neoliberalism as the dominant mode of social, political, economic, and cultural life. This potential calls for a reconsideration of capital as a transformative mode of social organization, whereby the definition of society is expanded to include the flourishing of nonhuman species, ecological cycles, and future lives.

Although concerns for socioeconomic injustice are a welcome contribution to the discourse on how to lower global GHG emissions, and I certainly embrace their inclusion in the agreement reached at the climate change talks held in Bali in December 2007, the variations of the "polluter pays" principle and the projected figures for increases in economic output and wealth as a result of opening up new markets in green products, services, industries, technologies, and energy still do not effectively address the underlying power relations produced by the neoliberal economy and the cultural norms reinforcing these relations. It was in opposition to the asymmetric power relations endemic to the global economy that representatives from low-income countries boycotted the Copenhagen climate talks in protest that their interests were being marginalized.

Neoliberalism can be described as a cultural mode of production that in turn defines the political economy. In this respect, I would go so far as to say that the principle of equity espoused by Shue and the introduction of carbon limits based on population size that Monbiot and Attfield propose sideline the power relations driving commodity culture and the consumer-credit economy because they invite a linear connection to be drawn between GHG emissions and the resources available to lower these emissions, all the while ignoring how capital accumulation is centralized.

Mass markets and aggressive moves toward commodity differentiation are what characterize today's economies of scope (which have gradually replaced economies of scale). Along with the rise in economies of scope, there has been a concomitant rise in carving out niche markets, one of which is the market in green commodities. A green commodity market includes recyclable or degradable materials, products that have a small ecological footprint, products whose ingredients have not been tested on animals, products that support socioeconomic development initiatives, and products that are produced with renewable resources. Such products range from clothing made from organic cotton to reusable organic cotton shopping bags, the hybrid car, recycled paper, biodegradable cleaning products, "natural" apparel (usually made out of hemp, organic cotton, or bamboo), "natural" cosmetic lines, and so on.

When green artifacts first appeared on the market in the latter part of the twentieth century, they offered an alternative to the excesses of the consumer global market and the individualism of commodity culture. This oppositional position made them popular, but it has also ironically increased the clout of the green commodity within the market, resulting in the seamless promotion of those selfsame relations of production (equating the liberal's goal for social justice with economic growth, understood as private enterprise freed of government interference) originally argued against. From here, the artifacts have been incorporated into the system of commodity production—hence, the slogan/title for Tamsin Blanchard's book *Green Is the New Black*![25]

The commodification of the environmental and social justice movements in green artifacts celebrates principles of sustainability by placing the energetic power of moral feeling that social and environmental responsibility gives rise to in the service of the free market. The will to do good characterizing the socially and environmentally conscious consumer that has enabled, for instance, the backlash against bottled water among consumers stems from the same energies that a corporation such as BP (which changed its name to "Beyond Petroleum") seeks to profit from with its new green brand image. And we didn't need to hold our breath for long to see how BP revitalized its brand after the 2010 *Deepwater Horizon* oil spill. If its advertising expenditure after the spill is any indication, its strategy was to turn the crisis into a branding opportunity. Between April 2010 and July 2010, BP spent in excess of $93 million on corporate advertising—three times its advertising budget for the same period the previous year.[26]

Among the initiatives the corporate oil giant uses to target and gain the support of this new socially and environmentally responsible consumer is investment in new green technologies and climate research. The corporation's sustainability mission statement emphasizes such initiatives:

> BP wants to be recognized as a great organization—competitively successful and a force for progress. We have a fundamental belief that we can make a difference in the world.

We help the world meet its growing need for heat, light and mobility. We strive to do that by producing energy that is affordable, secure and doesn't damage the environment.

BP is progressive, responsible, innovative and performance driven.[27]

We need to remain alert, however, to the reality that the connection between socially and environmentally responsible consumption patterns and corporate activities is not linear.

The idea of a green commodity market is an oxymoron because it treats climate change as another opportunity for neoliberal economic growth, placing the solutions to climate change once more in the service of the free market, privatization, and competition. A few natural- and green-product logical fallacies have popped up in the process: If there are currently no industry standards in the United States for "natural" cleaning products, what standard are companies such as Clorox using for their new Green Works product line? What if a company, such as Clorox, has a "natural" product line along with many other products that are incredibly toxic (such as bleach, which poisons the environment) or that contain ingredients tested on animals? Does that organic apple you are eating come from a farm that respects the rights of its workers or not? The point is that two wrongs do not make a right. What is needed is a much more nuanced approach to how wage labor, environmental care, capital production, and the political concepts of justice and equality connect.

Introducing a new breed of commodities into the market as a way to solve the climate crisis is just a displacement activity. To suggest that revolutionary subjectivity can be produced as a fashion statement—"Green is the new black!"—is abhorrent for the simple reason that it turns the collectivist impulse at the core of political action into mere narcissism and lands us squarely back in the logic of individualism again. Worse still, the squeaky clean image of green commodities masks the exploitative conditions of their production.

For a long time, the sustainability movement, as Paul Hawken has so lucidly described, occupied a marginalized sociocultural position in respect to commodity culture and the global free-market economy.[28] Today, however, the popular catchphrase and slogan *sustainability* is being seamlessly integrated into dominant modes of economic production and normative forms of political and civic life. The three levels of commitment defining the sustainability movement—social equality, indigenous rights, and environmental justice—have now entered a system of exchange that participates in the institutionalization of commodity culture and the free-market economy. This participation is co-opting the politics of the sustainability movement because it implicates the movement in the very system of relations that the movement struggles against, unleashing some unlikely marriages between hippies and the corporate suits. For example, in 2006 Colgate-Palmolive bought Tom of Maine's for $100 million, and in 2007 Burt's Bees was bought by Clorox for $913 million.[29]

A sticky conundrum is emerging here: sustainability needs to become popular in order to be ecologically useful, but the moment it does so, the more susceptible the movement's theories and practices are to being abused by those very entities that are ecologically harmful. If the popularization of the theories and practices of sustainability are predicated upon popular culture, then they also presuppose the selfsame neoliberal economic structures and cultural values that inform dominant relations of production—consumption and commodification. The popularization and concomitant commodification of the sustainability movement mark the institutionalization of sustainability discourse and, as the case of the corporate greenwash attests to, posit the moment when the principles and practices of sustainability are diluted.

My concern throughout this book is that the more green commodities and the image of a green lifestyle are turned into a fashion statement (basically a narcissistic gesture), the more the complex and dynamic political and ethical dimensions informing the politics of climate change are eclipsed. In this context, climate change, environmental degradation, and the inequities they produce are reduced to a self-referential fashion statement because they become raw material for the production of surplus-value. It is for this reason that the corporate sector sees the new ecochic movement as a marvelous opportunity to maintain a competitive edge in the global economy. Likewise, the more leaders of the sustainability movement proclaim that the only realistic way forward is to hop into bed with the free market, the more the formal structure of opposition constituting the movement is compromised.[30]

This brings us back to the problem of capitalizing from climate change. Green commodities have entered the market at a time when the struggle over the resources driving the free-market economy—water, food, land, energy—are being depleted. Although I remain suspicious over whether a new kind of commodity or the greening of neoliberal economics can effectively counter the unsustainable aspects of a commodity-driven culture, I am more concerned with the reification of material life that climate capitalism advances. Climate capitalism neutralizes the politics of climate and environmental change because it advances, reproduces, and reinforces the oppressive material economic conditions and structures endemic to commodity culture and the free-market economy—the selfsame system that has produced climate warming. And as I outline in the next chapter, politics as consumption (taking on a green lifestyle or buying green commodities) does not sufficiently challenge this system.

Green lifestyes do not challenge this system

2

GREEN ANGELS OR CARBON COWBOYS?

Three men are enjoying a sunny afternoon at the poolside of their luxurious retirement home when Bob asks his two friends what kind of work they did in the past that landed them in such a salubrious setting at this late stage of their lives. Joe explains that he had a corner store selling cold cuts; he did reasonably well until one day he turned up at work, and the place had been completely burned to the ground. "The insurance policy paid me a nice sum of money and set me up for retirement," he says. Then Jon chimes in: "I had a packing warehouse down by the river. Business was steady but nothing remarkable. One day, the place simply flooded. Insurance, you gotta love it." At which point, Bob shares his story: "I had a tailoring business in the middle of town—that is, until a hurricane swept through the area. Luckily I was insured!" The other two men stare incredulously at Bob and exclaim: "How do you make a hurricane happen!"

Regardless of how ostensibly amusing this joke is; it has a serious side to it, for in many respects it attests to the current situation we find ourselves in. Joe and Jon made a decision to destroy their business by calculating the costs and benefits of their actions. The risks of bankruptcy were offset with the certainty that the insurance company would compensate them for the destruction wreaked by fire and flood. Bob, however, was struck by a natural disaster and luckily had an insurance policy to help compensate for his losses. The first two adopted a cost–benefit strategy that sought to optimize the costs associated with risky—not to mention fraudulent—behavior. Bob's decision to take out an insurance policy was to strengthen his ability to absorb shocks and bolster resiliency in response to such shocks. He assessed possible situations that might occur in the future and took out an insurance policy as a way to safeguard himself and his assets; his assessment paid off.

The voluntary carbon-offset market is not that different to the situation posed by this oft cited Jewish joke. Both the offset market and the three retirees' insurance policies are voluntary investments against future risks, and they offer a significant challenge to the "act now, think later" premise that, as the global economic crisis at the beginning of the twenty-first century proved, pervades contemporary business decision-making processes. As mitigation strategies set out to lower the costs of climate change, the carbon-reduction market operates like an insurance policy against the uncertainties of climate change and its future effects. The questions that arise, though, are: How do we assess the mitigation strategies implemented? Are they evaluated purely in economic terms in the way that Joe and Jon's cost–benefit approach does?[1] Or do we need also to factor into the equation a community's adaptive capacity to absorb the shocks inflicted by climate change? Neither of these questions, however, addresses the underlying structures that lead to greater vulnerability in the context of climatic change—inequities, power relations, and poverty dynamics—the selfsame structures that produce an uneven geography through which the flows of capital continue to move largely uninterrupted. This problem is one that largely slips below the radar of the carbon-offset market and the criteria used to assess the usefulness of the carbon offset. Indeed, I argue here that the voluntary carbon-offset market merely facilitates and maintains asymmetric relations of power throughout the global economy, bolstering the economic and political influence of the powerful at the expense of the less than equal.

The safe limit for atmospheric CO2 is 350 ppm, and it has not stayed within that amount since 1988, with figures since then consistently exceeding it. The U.S. National Oceanic and Atmospheric Administration reports that average annual concentration of atmospheric CO_2 in 2010 reached 389.78 ppm.[2] For the global average temperature increase to remain below

2°C compared to preindustrial levels, and in an effort to make the situation of climate change and the problem of GHG emissions producing it more tractable, climate change policies and strategies have been introduced the world over. Out of this action has emerged a new carbon economy that sets out to facilitate the growth of a low-carbon economy that results in the least possible output of GHGs into the atmosphere, if not eventually a carbon neutral one.

In the carbon economy, GHGs are bought and sold with a view to reducing the total amount of carbon emitted.[3] The carbon economy operates on the neoliberal presumption that the market will sufficiently discipline polluters, all the while boosting the profit margins of those whose business practices engage environmental issues. Simply put, this market employs economic policy to abate emissions. The primary justification used for the effectiveness of the global carbon market to lower GHG emissions is the equivalence principle. Adopting a stringent environmental perspective, the equivalence principle operates on the premise that geography is irrelevant for emissions-reduction projects because GHGs uniformly mix in the atmosphere. The central idea is that the bad effects of carbon emitted in one part of the world can be neutralized by projects that sequester, reduce, or prevent carbon from entering the atmosphere in other parts of the world.

The precedent for the current carbon market lies with the United States, which developed a pollution-trading market in the 1980s to prevent acid rain. This economic initiative successfully reduced sulfur dioxide emissions that cause acid rain. It was later modified and fine-tuned to become the carbon-trading system developed as part of the Kyoto Protocol. Prior to the protocol, the UN Framework Convention on Climate Change, which was signed at the Earth Summit held in Rio de Janeiro in 1992, had established a series of nonbinding agreements to stabilize GHG emissions, assist developing countries in lowering their emissions through technological transfer, and protect the world's carbon sinks. The voluntary nature of the agreement proved insufficient, though, which prompted a compulsory agreement to be developed at the next meeting of world leaders held in 1997 in Kyoto, Japan: the Kyoto Protocol.[4] The protocol established binding GHG emissions targets (a 5.2 percent reduction in emissions by 2000 as compared with 1990 levels) for each of its member countries.

In order for the protocol to take effect fully, member countries must ratify the agreement (the United States and Australia have not ratified yet).

Countries that have ratified are called "Annex 1 countries." The protocol's first task was to assist Annex 1 countries in realizing the emissions reductions it set, so Flexible Mechanisms were instituted—Emissions Trading; Joint Implementation (JI); and a Clean Development Mechanism (CDM)—to establish a framework through which the trading of GHG emissions could occur.

The Kyoto Protocol's Emissions Trading scheme is outlined in Article 17.[5] It allows countries that have not used all their assigned amounts of emissions for the 2008–2012 commitment period to trade or sell those unused emissions to countries that have exceeded their emissions targets. In addition, emissions units such as a removal unit (i.e., reforestation activities), an emission-reduction unit (ERU), or a certified ERU (such as a clean-development mechanism) can also be traded and sold under the emissions-trading scheme established by the protocol.

The voluntary carbon-offset market has in large part emerged out of the two project-based mechanisms for reducing emissions offered by the protocol: the JI and the CDM. The JI permits developed countries (Annex 1 countries) to collaboratively invest in and develop emissions-reduction projects as a way to meet their emissions targets.[6] The carbon credits generated by JI are the ERUs. The benefit of the JI is that it is an economical and flexible way for a country to reduce its emissions.

The CDM is a sustainable development scheme that allows developed countries (Annex 1 countries) to invest in emissions-reducing projects in developing countries (Annex 2 countries). In addition to being project based, the CDM is a market-based approach to solving the problem of GHG emissions. It has three primary goals:[7] first, to reduce GHG emissions; second, to build developing nations' low-carbon technology capacity through a system of technology transfer; and third, to foster sustainable development. For example, instead of using high-polluting technologies or energy systems, developing countries such as Brazil, Chile, China, India, Indonesia, and Nicaragua are paid to use low-polluting ones.[8] The certified emissions reductions arising out of the emission-reduction project produce credits that can be bought or traded. One credit equals one ton of CO_2 equivalent gas. Selling the credits helps offset the higher outlay costs for low-polluting energy solutions.

The CDM approach is not only preemptive, but also a more cost effective way to lower global GHG emissions because the cost of transforming existing energy infrastructure in the developed world is more than the cost of installing a low-carbon energy infrastructure.[9] The idea behind the CDM is to provide a financial incentive for developing countries to become involved in lowering their GHG emissions. Further, wealthy countries can generate credits by subsidizing projects in developing countries that save or reduce emissions. A saving of more than 2.9 billion tons of CO_2 was projected for the first implementation period (2008–2012).[10]

All Kyoto carbon emissions–reducing projects have to meet the "additionality" requirement: the emissions savings have to be in addition to what would have otherwise occurred. For instance, project developers have to demonstrate that the emissions reductions or savings would not have happened without the project in question. The Designated Operational Entity is an authorized third party who certifies that the additionality requirement has been met and that the stated emissions reductions have in fact been achieved.

In addition to the Flexible Mechanisms offered by the Kyoto Protocol, another important institutionalized market for offsets was established in Chicago in 2003, when Chicago instituted a voluntary GHG and carbon-offset trading program, the Chicago Climate Exchange (CCX).[11] Since its inception, the CCX has reduced approximately 700 million metric tons of carbon, which equates with 140 million cars being taken off the road.[12] Companies voluntarily enter into a legally binding agreement to reduce their emissions. The CCX operates on the premise that companies have begun to factor climate change into their corporate governance agenda and as part of their corporate environmental responsibility portfolios. In order to meet their targets, companies can either purchase pollution offsets from other entities whose projects lower GHG emissions, or they can purchase emissions from other organizations that fall short of their emissions cap. As such, the CCX uses a version of the cap-and-trade system. It can paradoxically best be described as being "voluntarily mandatory."[13]

Not long after the CCX, the EU instituted its own program in 2005 to assist union countries to meet EU climate policy—the EU Emissions Trading Scheme (ETS). The EU has strong climate change legislation, and the driving force behind making these goals a reality is the ETS.[14] The EU ETS system is mandatory, and it boasts of being "the first and biggest international scheme for the trading of greenhouse gas emission allowances."[15] The scheme involves approximately eleven thousand power stations and industrial plants over thirty countries (the twenty-seven EU member states plus Iceland, Lichtenstein, and Norway). The EU ETS operates on a strict system of cap and trade: a limit is set on the amount of GHG emissions that industry (power plants, iron and steel works, oil refineries, combustion plants, manufacturing plants) can emit. An allowance of emissions is allocated, and depending on how many emissions are used, a shortfall can in turn be sold, or if more emissions are needed, they can be bought from another entity that has not depleted its emissions quota. In order for the allowances to retain their value, the EU ETS sets a limit on the number of allowances it distributes. Over time, the EU cuts the number of allowances that are issued so that overall emissions fall. The EU accordingly estimates that by 2020 emissions from the EU will be 21 percent lower than those in 2005.

Critics point to the overallocation of allowances by member states to parts of the energy sector that they have a vested interest in protecting. This overallocation has resulted in windfall profits for these organizations because they are able to sell their unused allowances.[16] During the first trading period (2005–2007), the price of carbon allowances fell to zero because the EU misjudged the number of allowances it would issue and even allocated allowances free of charge. In an effort to make the system more equitable and rigorous, the EU intends to start auctioning off allowances. In other words, the EU recognizes that government can use the market to help achieve its emissions targets, but the market cannot be left to its own devices; it also needs regulation.

In addition to the larger institutional carbon-market frameworks mentioned, a host of smaller independent entities participates in the carbon-offset market. Projects that reduce, sequester, or prevent CO2 emissions generate what is called a "carbon credit" that a consumer can purchase to offset a single or long-term activity that generates GHG emissions. A common one-off activity that consumers can purchase a carbon offset for is air travel. A more long-term activity might be offsetting annual car mileage. Offsets are sold to individuals, groups, or business entities to cancel or mitigate the GHG emissions they produce. One ton of carbon typically costs anywhere between $5 and $25. Included in this price are the cost to offset the carbon in a given program and other administrative costs such as salaries and transaction expenses.

There are two kinds of offset markets: voluntary and compliance. The voluntary market consists of consumers who are not obligated to lower their GHG emissions but choose to do so of their own accord. The compliance market consists of entities that are legally obligated not to exceed a certain number of GHG emissions; in this market, though, an entity can buy offsets in order to ensure that it does not surpass this limit. The carbon-offset market is made up of retailers who promote, advertise, and sell an offset; project developers who are involved with establishing an offset project or program; middlepersons who bundle offsets together and sell them; the brokers and the actual exchange trading in offsets (such as the CCX); and third-party quality-assurance officers or organizations that ensure the legitimacy of an offset—for instance, the World Wildlife Fund with its Gold Standard.

Offsets are typically purchased from a retailer or broker, but some people purchase their offsets directly by investing in a project. Common projects include investment in renewable power (wind, solar, or tidal energy schemes) or forest offset projects (afforestation, reforestation, and forest-management activities). A person can now even pay for professional services with carbon credits. For example, Cueto Law Group (Florida) allows clients to pay up to 20 percent of their legal fees with carbon credits. Many retailers now focus on the sale of offsets, and the number continues to grow despite the global economic downturn.[17]

Although quality-assurance standards do exist, it is not mandatory that offsets comply with them, which leads to the question of how a project's carbon offsets can be objectively and consistently measured and verified. Added to this problem are the limited amount of information that a consumer receives concerning the credibility of the offset purchased and the subsequent dilemma over the best way to measure the impact of offsets. All this brings us to the question of value, for there is no single regulatory body that oversees the assessment and validation of carbon-offset value, thereby making the offset commodity more vulnerable to the whims of a speculative market because it exists solely on the market as an empty signifier into which all the energies of good ecological feeling that come from the identity of being a carbon-neutral individual are seamlessly invested.

Yet even if the value of the offset can be reliably measured, it still would not make the offset commodity an inherent "good." The production and circulation of this kind of commodity, despite what the equivalence principle might encourage us to believe, do not take place in a vacuum. Some projects openly disregard political struggles already taking place in areas where carbon-offset programs are introduced. In these situations, the programs have the disastrous effect of putting climate change politics to work to make capitalist modes of production coherent at the expense of subordinate groups, whose own political struggles are in turn rendered invisible and inaudible. For these reasons, as we move forward to decarbonize societies the world over, we need to address not only questions concerning the integrity and effectiveness of voluntary carbon-offset projects, but also the more Marxist problem of the social relations of production that the offset commodity sustains and at times amplifies.

It can be argued that because the United States was able to lower its sulfur dioxide emissions successfully through the right-to-pollute market that came into effect at the end of the twentieth century, we have a solid precedent in favor of a carbon market. Why wouldn't this system work for carbon emissions? One serious hiccup in this argument, however, is the issue of scale and the coordination of a consistent set of standards and methodologies across the world. To deal with acid rain, the United States had to monitor only a few thousand smokestacks in one region (the Midwest), whereas the global market in carbon involves many more power plants and companies dispersed throughout different parts of the world in different sovereign territories.[18] Further, in order to be effective, carbon offsets and amelioration efforts require the coordinated effort of various governments, entrepreneurs, communities, and NGOs, each of which have different capacities to monitor, regulate, and enforce emissions standards, not to mention different capabilities to absorb the costs of enforcement. Aside from the absence of a consistent set of criteria that can be used to assess, verify, and certify carbon-emissions projects, there is a very real concern over the integrity of the carbon market.

Michael Wara has pointed out that the global carbon market established by the CDM is a market for all six of the Kyoto Protocol gases (CO_2, methane, nitrous oxide, hydrofluorocarbons, perfluorocarbons, sulfur hexafluoride), not just CO_2, despite the fact that the reduction of CO_2 is of most concern because it remains in the atmosphere one hundred years and is emitted in massive amounts at an accelerated rate. He indicates that the largest volume of credits from CDM projects arise from capturing and destroying trifluoromethane (HFC-23), a "greenhouse gas that is a by-product of the manufacture of refrigerant gas." These gases amount to 30 percent of CDM credits, accounting for €4.7 billion in credits for the first compliance period. Wara remarks, "In fact, HFC-23 emitters can earn almost twice as much from CDM credits as they can from selling refrigerant gases—by any measure a major distortion of the market." The reason for the distortion is cost. HFC-23 is a cheap non-CO_2 credit. It is cheap to cut HFC-23 emissions—so much so that in the industrialized world manufacturers voluntarily reduce their emissions this way. Yet, as Wara mentions, it would be cheaper to pay for the installation of technology that captures and destroys HFC-23 than paying CDM credits for HFC-23. He calculates that this option would save €4.6 billion in CDM credits.[19]

Although Wara recognizes that the CDM's goal is not just to lower emissions, but to do so in a cost-effective way, he does not problematize the CDM's free-market-based approach. He does not go on to discuss how the principles of the free market give rise to the very problem he is identifying. The CDM subsidizes the developing world to reduce their emissions; however, by combining market mechanisms with development and carbon-emissions reduction, it is the most lucrative, not necessarily the most cost-effective, way of generating credits that is being pursued. And why not? This is exactly what a free-market-based approach strives to do: accumulate capital. It is this neoliberal premise—that the free market will adequately solve climate change—at the core of the carbon-offset market that I interrogate further on, but first let us return to the question of assessment criteria.

Serious concerns persist over whether offset projects fully comply with the additionality requirement or are involved in double-counting (selling the same offset more than once). In large part, the reason for these concerns is that the value of carbon credits is difficult to validate. Offsets can be sold for projects that simply do not exist. For example, Vatican City encountered this situation when it was given a generous donation to the value of €100,000 in 2007 from the startup carbon-offset company Planktos/KlimaFa. The gift came in the form of carbon offsets to be generated from planting trees over fifteen hectares of land on an island by the Tisza River in Hungary. The offsets would have made Vatican City the first carbon-neutral state. In a highly publicized event, Russ George, president and CEO of Planktos/KlimaFa, launched the "Vatican forest" by presenting Cardinal Paul Poupard with a carbon-offset certificate in July 2007, thereby piggybacking the donation to rouse publicity for Planktos/KlimaFa. Even by 2011, however, the trees still did not exist. The Vatican went on to take legal action to protect its reputation.

The Vatican carbon-neutral debacle leads us to the issue of integrity. Planktos was forced to shut down in December 2007 after selling offsets for a geoengineering scheme that went terribly wrong. The project was to use iron fertilization for carbon sequestration by dispersing iron dust over 2.4 million acres of the South Pacific. The idea was simple: grow phytoplankton algae that would absorb CO_2 and then sink to the seabed. Russ George sold the offsets at $5 a ton. The scientific community raised doubts over the project, claiming that distributing large amounts of iron filings into the ocean might cause more damage than good.[20] Plankton blooms might actually result in an increase in GHGs because they can lead to the release of methane and nitrous oxide into the atmosphere.[21] As a consequence, the project came to screaming halt.[22]

Connected to the subject of integrity is the topic of temporality. There is an urgent need to develop a reliable method of quantification that engages with the temporal dimensions of carbon storage.[23] How long does a carbon-sequestration project need to be in effect before it reaches an equivalence with prevented emissions? The answer requires formulating an equivalence factor between sequestration and CO_2 emissions. In other words, how successful is a project at achieving its stated targets over time?

Measurement aside, there is a whole gamut of sociopolitical ramifications to carbon-offset projects because the logic of the free market underpinning the carbon-offset market obscures the material effects of the dynamic of production and consumption. Let us consider the example of the two hundred *adivasi* (tribal people) from the Dhule District of India who had worked the same piece of land for generations. The farmers were offered $4,000 in 2007 if they would allow windmills to be built on their land. The project was part of a renewable-energy carbon-offset scheme. The farmers refused, but the 550 windmills of the Suzlon Energy Ltd. wind project were installed regardless.[24] The situation constitutes nothing less than a massive land grab. It left the farmers without a livelihood, broke up the community (farmers had to leave the area to find work in other parts of the country), and put an end to the promise given to the farmers of legally owning the land in the future.[25]

In a report published by the World Rainforest Movement in December 2006, Chris Lang and Timothy Byakola describe the human costs of one offset project in Mount Elgon, Uganda.[26] The nonprofit organization Forests Absorbing Carbon Dioxide Emissions (FACE) teamed up with the Uganda Wildlife Authority (UWA) to plant twenty-five thousand hectares of trees in Mount Elgon National Park to absorb CO_2. The idea was to sell the offsets to the for-profit Dutch offset companies GreenSeat and Climate Neutral Group. The project began in 1994, but the land it was slated to take place on was wrought with a history of violence against the forest-dependent indigenous population, the Benet people.[27]

The FACE–UWA forest carbon-offset project on Mount Elgon was the outcome of an alliance between international aid agencies and the Ugandan government to "conserve and use sustainably the delicate mountain ecosystem." The project began regardless of the ongoing land disputes between the UWA and the Benet people, who had been residing in the forest since 1956. A research team from the Centre for Development Studies of the University of Wales warned that there was "a clear bias towards conservation rather than considering the needs, hopes and desires of the people who will be affected." The Protected Area Management and Sustainable Use Program, funded by the World Bank, provided financial support to the surveying of the Mount Elgon National Park boundaries—the selfsame boundaries at the heart of disputes between the UWA and the Benet people. The World Bank had been involved with capacity-building projects in Uganda since the 1980s and had funded a $38 million European Commission Natural Forest Management and Conservation Project there from 1990 to 1993 that had also resulted in the eviction of the forest dwellers (approximately 130,000) from their lands without compensation.

Uganda (handwritten margin note)

Uganda + violence (handwritten margin note)

In addition to securing and establishing Uganda's parks, the $50 million of the Protected Area project was used to "train approximately 1,300 rangers in paramilitary skills, build capacity of staff, demarcate parks and develop infrastructure." Evictions led by UWA staff, police, and soldiers from Uganda's People's Defense Force were violent to the extreme: people were killed, beaten, and tortured, and women were gang-raped. In the process of resettling the Benet people, the UWA set fire to their homes, destroyed their crops, and confiscated cattle that remained inside the "red line" (the UWA boundary line of red sinking markers). When queried about the level of brutality waged against the indigenous population, one UWA official explained: "Mount Elgon National Park is an international conservation area. So we have to protect it from destruction."

The evicted were left landless and homeless and were robbed of their livelihoods. Yet despite all the violations waged against the Benet and the violence surrounding their land-claim disputes, at the Rio+10 Earth Summit in August 2002 Alex Muhweezi, the Uganda International Union for Conservation of Nature representative, commended Uganda's forest-conservation program, reporting to journalists that "Mount Elgon had been degraded but had been re-planted with forests to absorb CO_2 emissions." Lang and Byakola, however, state that "in April 2004, SGS [Société Générale de Surveillance] carried out a surveillance visit to Mount Elgon, to assess whether the UWA–FACE tree planting project continued to meet FSC [Forest Stewardship Council] guidelines. SGS's public summary of the surveillance visit makes no mention of the conflicts surrounding the Mount Elgon National Park and makes no mention of the fact that the Benet were suing the government."

The point of describing these examples in detail is to bring into focus the different levels of violence embedded in the fabric of the carbon-offset commodity. As the individual or business buys offsets to counter the violence perpetrated against both the environment and future generations, they may also indirectly engage with another kind of violence, as demonstrated via the example of the Benet people in Uganda and the *adivasi* in India. And perhaps the difficulty arises with the consumer's terrain of political struggle, which is limited to the realm of immaterial exchange value, thereby placing the production of political subjectivities in the service of fixed and circulating capital flows. A brief overview of recent trends in the voluntary carbon-offset market illustrates this point further.

In 2008, the voluntary carbon market was estimated at $728.1 million, but in 2009 it was at $387.4 million, marking a drop in values of about 47 percent.[28] In order to shed light on the reasons behind the drop, it might be helpful to examine the conditions of the global economy over this two-year period. As the World Bank reported, global GDP experienced a "sharp growth deceleration" in 2008 and subsequently contracted in 2009.[29]

If a dip in the global economy prompts fewer people to offset their carbon footprint, doesn't this trend highlight the limits of a neoliberal approach to lowering GHG emissions? The choices of the individual energy consumer are not rationally motivated by environmental concerns; they are driven first and foremost by an investment of the energies and affects arising out of feeling like a good responsible citizen with a green carbon-neutral identity, but that feeling in turn finds expression through consumption and can, I should add, quickly change direction, such as when hard economic times shift it to anxiety over financial matters. The irony is that consumers buy offsets to neutralize the carbon footprint of their excessive consumption so that they can continue consuming. Put simply, consumption trumps environmental concerns. What the priority exposes is the political problem of consumer politics.

If political subjectivity is predetermined by identity politics, it does not go on to be liberating. It fails to be transformative. If political subjectivity is solely an expression of the commodity form and determined by the carbon-neutral identity that the offset commodity supposedly represents, then politics never amounts to anything more than a displacement activity. It might be helpful to revisit an earlier point regarding the value of carbon commodities using Marx's discussion of money to scrutinize the logic of consumer politics driving the carbon-offset market and the depoliticizing effect that such a consumer-based construction of "political" subjectivity engenders.

In the opening of *Capital*, Marx defines value as a social relation:

> Not an atom of matter enters into the objectivity of commodities as values; in this it is the direct opposite of the coarsely sensuous objectivity of commodities as physical objects. We may twist and turn a single commodity as we wish; it remains impossible to grasp it as a thing possessing value. However, let us remember that commodities possess an objective character as values only in so far as they are all expressions of an identical social substance, human labor, that their objective character as values is therefore purely social. From this it follows self-evidently that it can only appear in the social relation between commodity and commodity.[30]

In many respects, the carbon commodity can be described as the commodity par excellence. In volume 1 of *Capital*, Marx explains that the value of a commodity is invisible; it is made perceptible only by the phenomenal form of money, whereby the value of the commodity is "expressed by an imaginary quantity" through the ideal form of money (price). The circulation process of commodities and money also presupposes the logic of property (the commodity owner who sells the commodity in return for money), wherein money is the "expression of the circulation of commodities."[31] Whereas commodities can fall out of circulation, as when they are consumed, money does not. As the medium of circulation, money necessarily remains in circulation. As an invisible entity, carbon is the pure form of Marx's concept of "value."

It would seem that, on the one hand, the carbon offset represents the perfect commodity insofar as its use-value will never become redundant in the context of climate change. It is also the ideal commodity on the global market because carbon emissions impact every person and all life forms both today and in the future. People therefore have a vested interest in ensuring that emissions are lowered so that the economy become carbon neutral. Such deep-vested interests are themselves a productive energy, which when placed in the service of market forces drive the value of the carbon-offset commodity. On the other hand, as a "market," the trade in carbon needs to grow—all with the aim of slowing down the supply chain feeding that growth. The paradox is peculiar and runs contrary to the logic of growth driving capital. However, one way to make sense of this situation is by using Marx's observations on commodity fetishism.

Marx writes: "A commodity appears at first sight an extremely obvious, trivial thing. But its analysis brings out that it is a very strange thing, abounding in metaphysical subtleties and theological niceties."[32] This statement can be compared to Michael Jenkins and Ricardo Bayon's argument that the mystical powers of the market be put to work so that people internalize the economic value of the environment,[33] proving Marx's point that the workers have internalized capitalist ideologies, making it hard for them to break free of the very system that incarcerates them. The subtlety Marx refers to is commodity fetishism. That is, commodities represent both material and metaphysical processes. The fetish prevents a person from appreciating the true environmental and social costs of a given commodity because the commodity is understood in terms of its money equivalent, which gains its power from the investments people make in it (carbon offsets make people feel green regardless of how much they consume and despite the material conditions of exploitation that go into the production of some carbon offset commodities).

For Marx, the fetish is fundamentally an inaccurate representation of the process of production. False consciousness produces the commodity fetish, thereby making the fetish an effect of false consciousness.[34] Put differently, the law of the market produces a fiction (the price of an invisible commodity—carbon). We identify with the plight of life on earth through the free market, which mediates our relationship to each other and to the environment in which we live. The free market is therefore simultaneously a mechanism of repression (replacing the investment of political energies in liberating change with an investment in free-market forces), a displacement activity (the benign image of political activity), and a repressive system that we unreflexively obey.

David Harvey has pointed out the political limitations of Marx's theory in this regard, remarking that Marx was not able to "create a space in his own thought in which the subjective lived experience of the working class could play out its proper role," and for this reason "he could not . . . solve the problem of political consciousness." Politics as consumption threatens neither the "adaptive powers of capital" nor the "processes of competition in particular," and, worse still, it facilitates the "mutual disciplining effect of the law of value in exchange and within production," which, Harvey points out, so many of Marx's critics ignore.[35]

Above all, from the logic of economistic struggle that Harvey provides, the fetishisms enveloping the carbon-offset market (offset commodities, the price fluctuations, consumers' carbon-neutral identity, and so on) "prevent any automatic translation of the experience [of political struggle] . . . into more general states of political consciousness."[36] As Larry Lohmann of the U.K. research group Corner House has commented, "Most governments, whether north or south, know by now that every year the world burns up 400 years worth of accumulated biological matter in the form of oil, coal, and gas. They are aware that the biosphere can't reabsorb all this carbon. They realize that an equitable way must be found of leaving most remaining fossil fuels in the ground."[37]

The voluntary carbon-offset market thus displaces the more pressing and pertinent question of how to end the current dependence on fossil fuels and, I should add to Lohmann's point, the problem of unsound land-use patterns (converting land with abundant ecosystems into agricultural or built environments and in the process damaging biodiversity and releasing ancient carbon from the earth's carbon sinks). Stressing the importance of finding a way to keep fossil fuels in the ground, Lohmann also demands that whatever solutions we come up with need to be equitable also.

Not only are the material realities of the unequal all too often rendered invisible by the logic of carbon commodity production, but their plight is often depoliticized as it is re-represented as part of a larger, "more important" environmental movement that in a well-meaning way aspires to save future generations and life on earth as we currently know it. This process of re-representation dehistoricizes and depoliticizes both the struggles of subordinate groups and climate change politics as well.

Under the benevolent guise of combating climate change, the discourse of climate change politics eclipses the struggles of the less than equal, as they vie for representation in the political processes that affect their everyday lives, while it finds expression in the free-market and commodity form. The re-representation of their struggles purely in terms of climate mitigation (a process that initiates a double silencing) is the effect of the centralization of capital through a unified global free market and the globalization of production that views carbon offsets as geographically neutral.

For these reasons, the voluntary carbon-offset market is a depoliticized gesture at solving the political problems arising out of GHG emissions and climate change. Its primary object is not to transform the conditions of a capitalist economy (exploitation and subordination) but to facilitate the production of a deeply depoliticized "political" subjectivity. Like the off-set produced by the market, the offset consumer's carbon-neutral political identity mediates and at worst obfuscates the violence of the social relations endemic to the production of carbon pollution and its subsequent transformation into a commodity. The question that remains is: Why are we so willing to define ourselves, our relationships to one another, and the environment in which we live through a capitalist system of appropriation?

The trade in carbon emissions facilitates the formation of a depoliticized subjectivity under the guise of a political identity. That depoliticized subjectivity is the effect of turning the criticisms of capital into yet another mode of capitalist production, simply enabling capital to incorporate the criticisms waged against it and in the process to defuse the political time bomb that climate change presents for business as usual.

The entire system of voluntary carbon offsets operates on the neoliberal premise that the market is the best way to solve environmental objectives. In what seems to be a perversion of the anxiety around the buildup of GHG emissions, GHGs have become an asset. To the extent that capital has utilized the "crisis" of climate change for its own objective—capital accumulation—the critique that environmental and social justice activists have leveled against global capitalism and the corporate sector has been neatly turned to the advantage of the global free market. Using the equivalence principle to frame the problem of GHG emissions and on this basis presuming that the projects of the voluntary offset market are somehow immune to the geographies of poverty, power, and inequality that underpin the global economy are fundamentally dishonest.

The voluntary carbon-offset market obscures the structures of exploitation and subordination endemic to the circuitry of capital as it flows from the back pocket of Rabobank to Suzlon Energy Ltd., but *not* to the *adivasi* farmers in India whose land was used without their permission for carbon offsets. The same structures can be seen as capital flows from the individuals and businesses buying offsets from Green Seat to balance out their air travel emissions to FACE, which owns the carbon-sequestration rights to the twenty-five thousand hectares of trees in the Mount Elgon National Forest in Uganda that provides Green Seat with carbon offsets to sell, as well as to the World Bank and the European Commission, but *not* to the Benet people, who not only lost their livelihoods and land but were also the victims of arson, rape, torture, and beatings as they were forced from their forest homes at the hands of UWA staff, police, and soldiers. From this standpoint, the voluntary carbon-offset market facilitates the expansion of power and extends the authority of the already influential at the expense of the vulnerable.

We urgently need to solve the problem of GHG emissions, but turning carbon into a commodity is not a solution; it is a cop out. All of us— Americans, Europeans, the *adivasi*, the Benet, the animals who are losing their habitats, workers, the ecosystems choking from pollutants spewed into the atmosphere, future generations—are in it together, and all of us are subjected to capital. It is the political subjectivity that arises from this realization of a shared experience and the will to change this situation of subjection that buying carbon offsets fragments and displaces. The problem with the voluntary carbon-offset market is not just that it provides a green light to citizens of wealthy nations to continue unsustainable levels of consumption, but that it also facilitates the violence produced by global capitalism in which all of us, at least those of us not crippled by poverty, are implicated.[38]

3

POPULATION

Developing countries are adding over 80 million to the population every year and the poorest of those countries are adding 20 million, exacerbating poverty and threatening the environment.

—Bill Butz, president of the Population Reference Bureau, 2010[1]

Population growth in news

The more climate change scenarios gain traction throughout the popular imaginary and concerns over environmental degradation mount, the more human population growth is becoming the subject of agitated discussion and debate, as much in the popular media as in scholarly circles.[2] The debate centers around how many people the earth can support and the ways in which population numbers drive changes in climate. The argument is that the more people there are on the planet, the more GHG emissions will be emitted and the more the earth's limited resources will be consumed. Surviving the potential health and environmental effects of climate change in large part depends on our ability to adapt and offset the severity of climate change. Because human activities are responsible for the large amounts of GHG emissions that have led to changes in climate, mitigating population growth has become entwined with the political discourse of climate change adaptation and measures to slow it.

In 2011, global human population hit the 7 billion mark. UN demographers estimate that between 2009 and 2050 the world's population will increase by 2.3 billion people and will peak at a little more than 9 billion. Most growth is projected to occur in the developing world, which will reach

7.9 billion people in 2050.[3] The reasons for the growing number of human beings are many: better health, improved average life expectancy, lower mortality and fertility rates, and increasing urbanization and economic growth throughout poorer regions of the world. Food, water, land, and energy resources are being placed under enormous stress, and many predict that these problems will become acute if the situation of population growth goes unchecked.[4] The IPCC "Special Report on Emission Scenarios" lists population growth alongside economic growth, technological change, and changes in patterns of land and energy use as a key factor driving GHG emissions.[5]

Noted climate scientist James Lovelock explains with his theory of Gaia that the earth is a dynamic yet stable living entity. Put differently, the earth is a single self-regulating physiological system of animate and inanimate matter. Human activities such as deforestation, emission of GHGs, overfishing of oceans, and so on are disrupting the dynamic equilibrium of the earth, thereby destabilizing the environmental conditions that have been favorable for life on earth as we know it to flourish. The two main culprits Lovelock identifies are fossil fuels and human population growth.[6] For this reason, Lovelock rallies on the side of efficiently lowering the population growth rate through the introduction of voluntary population controls. He maintains that the solution to overpopulation lies with women. He explains that "when women are given a fair chance to develop their potential they choose voluntarily to be less fecund."[7] Yet as a climate scientist Lovelock does not take into account the thornier politics surrounding reproductive policies and programs.

The social and health implications that arose as a consequence of China's one-child policy, instituted in 1979, are now widely known. One of the primary goals of the one-child policy was to slow dramatically China's population growth so that by the year 2000 the national population would be 1.2 billion people. If assessed purely on the basis of population numbers, the policy was reasonably successful. In 2000, China's population was 1.25 billion people, the government estimating that the policy prevented between 250 and 300 million births. Yet at what cost was this goal achieved? Chinese culture and custom favor boys more than girls because boys carry on the family name and are responsible for looking after their parents in old age (they act as a safety net). For this reason, rural couples whose first child is a girl have been allowed to have another child.[8]

The Chinese preference for boy babies unsurprisingly led to a skewed sex ratio. With couples selecting to abort female babies, the number of male births compared to female births increased. Female infant and child mortality rates grew as girls were neglected. Girls were less aggressively treated when they fell ill than their brothers were. In conjunction with this issue, the one-child policy has had broader negative social implications. The scarcity of women has fueled the commercial sex industry along with the illegal trade in abduction and trafficking of women for marriage. Medical scholars have flagged the potential impacts these activities have for the spread of sexually transmitted diseases and HIV/AIDS.[9]

Interestingly, India, which does not have national population controls, also has a distorted ratio of boys to girls, which is second to that of China. Female infanticide is common in India because a female child is a financial liability for the family, which has to save for her dowry, whereas a boy can be the source of future income because he is paid a dowry. Despite the fact that dowries were made illegal in 1961, the cultural tradition remains intact. As of 2010, the ratio of boys to girls in India was 1,000 to 914.[10] As in China, the reason for India's skewed sex population ratio is that the connection among reproduction, culture, society, and economics is complex, and oversimplifying this connection leads to socially unjust solutions.

State family-planning initiatives that aim to lower the birth rate efficiently by regulating the number, timing, and spacing of births are notoriously unjust.[11] Betsy Hartmann declares in her groundbreaking analysis of the politics of population control, *Reproductive Rights and Wrongs*, that the population–environment discourse informing family-planning policy in the global South and developing East leads to nothing less than "neofascist" population environmentalism. She demonstrates how efficiency-based programs are intimidating and restrict women's choices. These programs aspire to achieve numerical targets (often calculated on the basis of number of births or the number of women using contraception), adopt a cafeteria approach in poor communities by presenting women with several contraceptive options and often "encouraging" certain choices (IUDs, sterilization, the contraceptive pill, or implants such as Norplant), or offer programs that use incentives such as rewarding those women who accept contraceptives or who agree to being sterilized.[12] In the context of poverty, family-planning "incentives" that offer food and cash do not enhance a woman's reproductive choices; they are purely and simply coercive, for there is no "free" choice when you and your family are starving.

With regard to starvation in relationship to overpopulation, Paul Ehrlich and Anne Ehrlich have argued that the human race is procreating itself to death.[13] They use energy as a standard with which to measure the environmental damage of human activities. The picture they create directly challenges the trend to hold poor women of low- and middle-income countries responsible for the depletion of the world's natural resources and changes in climate. The Ehrlich equation has been used to calculate the environmental impact of population: $I = PAT$, where I is environmental impact, P is the number of people, A is per capita consumption, and T is environmental damage. The Ehrlichs later introduced a surrogate formula, $I = P \times Epc$, where Epc refers to per capita energy consumption. The surrogate equation provides a more nuanced analysis of the conjunction of population growth and environmental damage because it takes into account the dramatic differences in consumption between the developed and developing world. As a result, the authors reverse the standard conclusion that the rate of reproduction by poor women in the developing world is responsible for global overpopulation. In fact, when the variable of per capita energy consumption is incorporated into the population equation, the result is that the developed world, not the developing world, is shown to be the most overpopulated. Flattening the population–climate discourse to a question of numbers has had the unfortunate consequence of holding women in developing countries culpable for climate change and environmental degradation, a distortion that some statisticians have sought to rectify. For instance, Paul Murtaugh and Michael Schlax calculate the carbon legacy of individuals who have a child, factoring into their assessment where a person lives.[14] Their findings show a vast difference between the legacy of a U.S. female who has a child and that of a Bangladeshi female, with the American woman's legacy (18,500 tons) being two orders of magnitude greater than that of the Bangladeshi female (136 tons). As Murtaugh and Schlax assess the weighted emissions of descendents, they "explore the effects of individual reproductive behavior by tracing a single female's genetic contribution to future generations and weighing her descendant's impacts by their relatedness to her."[15] Although they recognize that a woman from the United States has a greater carbon legacy than a woman from China, they also add that by virtue of China's large population size its total emissions surpass those of the United States. Hence, we come full circle: the finger of blame once again falls on the reproductive labor of poor women in low- and middle-income countries.

I do not mean to suggest that the intersection of population and climate change be ignored; I am, however, proposing that the intersection be more

vigorously put to work. The discourse needs to welcome the critical input
of those working on the ground along with those whose research amplifies,
as opposed to simplifies, the social, economic, and political costs and risks
associated with population-control policies. Criticisms destabilizing the
work of family-planning interventions in low- and middle-income countries have led to frustrations such as those expressed by John Bongaarts,
Brian O'Neill, and Stuart Gaffin. Although they acknowledge consumption accounts for a big chunk of the climate pie, they also castigate "human
rights advocates and women's groups" for their criticisms of coercive population controls, accusing them of playing a "part in limiting the discussion
of the relationship between population and environmental change at the
1992 Earth Summit."[16] This kind of rejection of constructive and informed
criticism reinforces the structural obstacles that disadvantage poor women
the world over because it builds a firewall between the social impacts of
population-control practices and the environmental politics informing population-control policies. Although Hartmann's research does the
important job of dismantling the firewall, I would suggest that it has the
effect of a displacement activity, and, for this reason I believe that if we
intend to disassemble the barrier, we need first to defuse the social energies
and forces that built the firewall in the first place.

The anxieties over the growing populations in low- and middle-income
countries and the projected burdens this growth will pose for mitigating
and adapting to climate change does not neatly square with the geopolitical realities of the global economy. If we briefly return to a point I made
in chapter 1 concerning China's carbon emissions—wealthy countries
outsource their dirty manufacturing to China—we might be able to recognize the paradox in operation with regard to population. Many people
from low- and middle-income countries work in the service industry and
in the manufacturing of commodities that meet the consumption needs of
wealthy nations. In light of this situation, framing climate change as a problem exacerbated by the growing populations of low- and middle-income
countries quite simply shifts responsibility and culpability for climate
change away from affluent high-carbon societies.

So how is it that the population–climate discourse came to be defined
by the reproductive choices of individuals and more specifically the reproductive choices of poor women from low- and middle-income countries,
despite research that points to the high emissions arising from the consumption rates and lifestyle choices of the affluent? The answer to this

Finite Population *AR argument* *Hardin* *Children as assets in poor families*

question lies in part in the influential 1968 essay "The Tragedy of the Commons" by Garrett Hardin,[17] whose work carries a strong Malthusian aftertaste. In it, Hardin argued that despite advances in science and technology (genetically modified crops, new agricultural technologies, and so on), the world can support only a finite population because it is biophysically constrained, a position that Lovelock also puts forward. In Hardin's view, the problem of population growth arises out of human beings' self-interested nature. Although the earth's resources are held in common—they are collectively shared—individuals exploit those resources for their own benefit. Hardin maintained that individual rational behavior is characterized by self-interest, and the result is that the individual acts to maximize his or her own benefits even when the costs are collectively incurred.[18] He assumes that the commons dilemma is one of open access—people openly use the earth's resources without weighing up costs and benefits to others, instead of considering perhaps that they need to restrict the overexploitation of the commons. He was clearly in favor of a utilitarian approach to solving his interpretation of the ecological and social costs of population growth, arguing that the best way to achieve the greatest happiness for the greatest number of people is to limit individual freedom by placing constraints on the number of offspring women can produce.

Although Hartmann supports the position that people act out of self-interest when they decide to have several children, contrary to Hardin she looks to the conditions that constrain them to do otherwise. She criticizes Hardin's lifeboat ethics, which likens the earth to a "lifeboat in which there is not enough food to go around" and which posits the solution as "not to let the poor and the starving on board."[19] In response, she places reproduction in a broader social, political, economic, and cultural context, clarifying that children are like an asset in poor families. They contribute to the household either by babysitting their younger siblings while their parents work or by working themselves (girls often help collect water or firewood and boys help in the field). Adult children can also take care of their parents when they fall ill. And given the higher infant and child mortality rates that characterize poorer communities (largely due to poor mother or child nutrition or both), parents often have several children to ensure that some survive through to adulthood, at which time they become a safety net for their parents in old age.

Hartmann supports a humanitarian approach to population control, one that starts by challenging inequity and confronting the poverty that arises from this. Her position echoes that of Victoria Tauli-Corpuz, founder and executive director of the Tebtebba Foundation (Indigenous Peoples' International Center for Policy Research and Education). Tauli-Corpuz has publicly criticized those who view population growth as the primary cause of climate change: "The main thing is really the lifestyles— the economic development model that's being pushed. . . . [I]f you think population is the problem, and undertake centralized ways of controlling population growth, we will be in an even greater mess."[20] In place of the reproductive rights model, she advocates a holistic approach to reproductive health, clarifying that the quality of women's health is directly proportional to the opportunities and education available to them.

The direct correlation between the number of people, on the one hand, and climate change and environmental depletion, on the other, fails to account for the vast differences in patterns of consumption, which in turn correlate with the unequal distribution of power and resources around the world.[21] Although some analyses reflect distortions in patterns of consumption across populations, and others recognize the cumulative energy use and emissions effects of individual reproductive choices, all these analyses suffer from one significant blind spot: the individual subject is used as the primary unit of analysis.[22] By considering the complex ways in which environment, demography, time, and social resources interact, the notion of an undifferentiated subject is problematized.

Leiwen Jiang and Karen Hardee take an important step in this direction as they underscore the limitations of using the individual person's consumption patterns as the primary unit of assessment in developing climate change scenarios. They stress that the demographic component used in integrated assessment models assumes that "each individual in a population shares the same productive and consumptive behavior." They warn that this assumption leads to serious inaccuracies. In response, they recommend that the individual consumption unit be replaced with household size, arguing that the latter is a more reliable variable because the per capita energy consumption of a smaller household is much higher than that of a larger household. They also draw attention to the ways in which consumption patterns change across different kinds of households. For instance, aging and urbanization trends generally drive carbon emissions down.[23]

Recent research such as Jiang and Hardee's raises an intriguing problem for contemporary social and political theory. The liberal notion of the unencumbered individual who is free to express himself or herself and to

(margin handwriting: women's bodies to be regulated for "rights")

make choices in the world not only distorts projections for how population impacts climate, but also turns the bodies of poor women in low- and middle-income countries into a political site. Using climate change predictions to push for caps on population growth transforms the meaning of both the environmental and the social issues at stake. As the impoverished reproductive body of the poor becomes the subject of "democratic" discourse (in UN reports, popular media, government policy, NGO programs, allocation of funding, and scientific scrutiny), it is inserted into the public domain.

The dominant discourse ostensibly claims that the reproductive body of poor women needs to be regulated if the fundamental principles of human rights and individual liberty that a liberal modern democracy upholds are to be defended and if life on earth is to be secured.

Something interesting subsequently happens: as the reproductive coding of the bodies of poor women is deterritorialized, those bodies are seamlessly shifted into a liminal space between reproductive and productive labor. By recognizing this shift, I am not intending to valorize as subversive the in-between spaces in which they hover. As Brian Massumi has aptly noted, when theory attempts to overcome the hierarchies implied within dualistic thinking, it too often averts to the in-between spaces of hybridity and bordering in an "effort to find a place for social change again," the effect of which is to relegate social change to the site of marginality, which is in itself "defined less by location than [by] the evanescence of a momentary parodic rupture." Massumi incisively remarks that from here the question of how subversion can "react back on the positionalities of departure in a way that might enduringly change *them* becomes an insoluble problem."[24] In the absence of radically changing how the positionalities on either end of a divide work, the potentially antagonistic character of their relationship is neutralized.

The political problem is less one of how to release a body from occupying an in-between space than one of changing the oppressive positionalities that arise from either side of the divide in which the body is situated. Although the reproductive rights movement might aspire to emancipate women from reproductive labor, as Hartmann's and Tauli-Corpuz's work shows, a social and culturally sensitive approach has to address the different ways in which motherhood and reproductive and sexual health are connected to educational and work opportunities. Providing women with the technologies to decide when they will have children is not an inherently emancipatory move if all it means is that the laboring poor female body is

Moving women from kitchen → factory meaningless

then put to work in a context where she has no worker's rights. Rosi Braidotti aptly states: "All technologies have a strong 'bio-power' effect, in that they affect bodies and immerse them in social relations of power, inclusion, and exclusion."[25] This brings the analysis full circle because it means that framing the connection between climate change and population growth as a question of how to achieve reproductive rights is misplaced.

Another pertinent issue is how the reproductive rights movement connects the larger issue of achieving gender justice to the population debates of climate change politics. The reproductive rights approach to population control views its work as redistributing resources by providing women with the resources they need to make a choice over whether they will have a baby or not. It also recognizes that the definition of women in terms of their reproductive labor creates the marginalized social status women occupy, which in turn leads to gender discrimination. Last, it aspires to increase gender parity by providing women with greater access to the reproductive decision-making process.[26] However, cracks start appearing when capital infiltrates the reproductive rights movement. This problem is not a normative one; it is a machinic question of how the aims of reproductive rights activists are put to work.

For instance, one of the mechanisms through which reproductive and productive labor socially subordinates the bodies of poor women is the placement of the body's labor power in the service of capital accumulation. The emancipatory potential of the lived body is regulated as it is inscribed through a socioeconomic discourse of capitalist value. For instance, in the context of women's reproductive rights, reiteration reinstates the authority not of the "original" meaning (the reproductive body), but of sociopolitical processes that characterize capitalist production and exploitation. By the same token, the iterative process resignifies the reproductive meaning of the bodies of poor women from low- and middle-income countries as *liberated* (or potentially liberated) by the principles of the liberal democratic state (understood as a secular institution committed to the advancement of individual human rights), and in so doing it locks the affective potential of those bodies between reproductive and productive labor, neither of which changes how labor power is abstracted as part of the larger machinations of capital.

The issue is not just that the meaning of a body changes (passive to active subject) through reiteration, as Jacques Derrida claims in his use of the concept; it is that in the relations that situate and demarcate the *potentia* (creative power) of bodies in a "liberated" zone, the liveliness of labor power is abstracted and placed in the service of capital. It is therefore not the changed meaning of the body that matters as much as the machinic question of how the body's affective potential is put to work as part of reproductive and productive labor power. This is why Derrida's concept of iteration as transformative is depoliticized, not for the usual reasons Derrida is criticized—namely, that the deconstructive project of decentralizing truth values is nihilistic—for such a criticism oversimplifies the Derridean project, which takes care to emphasize that releasing the restraints of absolute Truth is in fact a liberating exercise. Rather, the central role Derrida accords the relational condition of meaning as constituted through the iteration of differences displaces the machinic question of how those relations work.[27] Despite the reproductive use of a body to produce the labor force (laboring in pregnancy, childbirth, childrearing, and domestic work) that is being transformed by reproductive technologies (Norplant, IUDs, and so on), the change that ensues when reproductive labor and technology interact in this way puts that same body to use as productive labor power working on the sweatshop floor.

In *Empire*, Michael Hardt and Antonio Negri envision a liberating role for *potentia* and immaterial labor, which they understand as a self-creating, mobile, and deskilled labor force.[28] In the later *Commonwealth*, they modify this position, commenting on the complicity between *potentia* and the biopolitical character of capitalist production. They unpack the technical composition of labor and capitalist forms of exploitation, describing three trends characteristic of the biopolitics of capital: immaterialization, feminization, and migration. Their definition of immaterial labor includes "images, information, knowledge, affects, codes, and social relationships." They add: "Living beings as fixed capital are at the center of this transformation, and the production of forms of life is becoming the basis of added value."[29] The second trend they identify is the feminization of labor, which they take to mean both the quantitative increase in the number of women entering the wage labor market and the qualitative changes in labor. For example, the work-day unit is no longer so clearly defined by the eight-hour working day; it has been turned into a flexible period, including part-time, contract-based, and informal forms of employment. This temporal smudging of work time and life trails alongside the blurring of productive and reproductive labor. It is also characteristic of what Hardt and Negri describe as "feminized" labor, in which the labor force has become infused with the qualities traditionally associated with work done by women, such as caregiving. The third labor trend is new migratory patterns as more and more women from low- and middle-income countries enter low-skilled, labor-intensive work in addition to jobs traditionally taken by women, such as those in the service sector (as nannies, cleaners, sex workers, and nurses).

By maintaining the Marxist dialectic of struggle, Hardt and Negri revitalize what Massumi describes as the logical consistency of the in-between—not as a "middling being but rather [as] the being *of* the middle—the being of a relation."[30] Unlike the liminal in-between discussed previously, the ontological conception of the "being *of* the middle" is conditioned by relationality. Indeed, as Marx identified, the antagonistic nature of this relation is where the potential for political change resides. Hardt and Negri state: "Capitalist production . . . is becoming biopolitical."[31]

Hardt and Negri understand the biopolitical dimension of capital to refer to the coupling of commodity production with the production of social life, and the point they make here reminds me of the women I met who were living and working in Dharavi, Mumbai, one of the world's largest slums. I was struck by just how much the emancipatory promise of immaterial labor had not been met.[32] Thus, for these women the regime of appropriation and enclosure through which the biopolitical mode of production operates (under the guise of reproductive rights) is also another expression of the division of labor. It is not that different from the caste system, guilds, or modern industrialization.[33] So although the development of biotechnologies is an indispensable feature of the biopolitical mode of production, as Hardt and Negri outline, for the women of Dharavi the antagonism between labor and capital is still a constitutive political struggle. The struggle over how reproductive technologies are put to work is key here because just as quickly as these women's bodies are liberated from reproductive labor, they are inserted into the system of productive labor. This situation cuts to the heart of reproductive technologies and the debate over population control as it is tied to climate change politics. The *freed* bodies of the women in Dharavi have in effect become the structural source of their exploitation. Having fewer babies, they now have more time to work sitting on a dirt floor sorting through the trash the wealthy left behind, without protection, breaks, sick leave, or workers' compensation.

Reproductive technologies are basically the machinery through which the biopolitical mode of production appropriates the bodies of poor women so that more of the force and energy of material life can be appropriated and enclosed. The more we push population into the foreground, the more consumption rates and the extraction of resources and environmental exploitation needed to sustain commodity production get pushed into the background. Are the poor women from low- and middle-income countries having fewer babies so that the affluent can continue to consume a steady line of cheap commodities that are made by the cheap labor of those selfsame women—for instance, those who have migrated from rural India and now work in the slums of Mumbai because they have more time to do so now that they have fewer children to take care of?

The point is that both reproductive labor and productive labor carry out the same economic function: capital accumulation. From this perspective, hovering in the liminal space between productive and reproductive, labor is itself a relation of production. As Gilles Deleuze and Félix Guattari explain, capitalism works to "decode and deterritorialize the flows" of bodies.[34] Simply put, following close behind the deterritorialization of the reproductive determination of bodies is the reterritorialization process waged by capital. As climate justice is presented through the lens of reproductive rights, the bodies of poor women from low- and middle-income countries are reconfigured through the axiomatic of capital's fulfilling all the three labor trends that Hardt and Negri so aptly identify.

Here we return to the women I visited and spent time listening to in both rural India and the slums of Mumbai at the beginning of 2011. They left me wondering how they experienced the liberation that came from access to family-planning programs. It seemed to me that what this "liberation" had facilitated was merely a shift in location: the same institutionalized power relations defined how their bodies were situated and organized within the social field. That is, the women of the slums may have moved out of the private patriarchal world of the family in rural India, but they were quickly inserted into the private sweatshops of the plastic-recycling center of Dharavi earning 100 rupees a day (approximately $2.25) and returning home each day to a list of domestic chores in a tiny ten-by-ten-foot room/ home that had running water for three hours a day and no sewage system. I was left with no doubt as to these women's creative potential to transform their lives as workingwomen. They proudly spoke of the respect they had gained in the eyes of their husbands because they, too, had become breadwinners, yet I also had no illusions over what this kind of *potentia* (creative power) could achieve.

The interface of climate change and population-control discourses operates by politicizing the female body. Whereas Hartmann stresses the manner in which population programs are geared toward regulating the bodies of poor women, I would maintain that the source of the problem is less the despotic machine that codes and recodes women's reproductive function and bodies than the ways in which such programs simply facilitate the freeing up of those bodies so that they are better situated to release the flows of labor and capital.

As the hot-button issues of climate change and environmental degradation are tied to population growth, the knee-jerk policy response has been largely utilitarian: limit individual freedom (to reproduce) so as to maximize the freedom of the collective (quality of life). Ironically, the goal of this argument is the source of the problem. In a neoliberal capitalist context, "quality of life" is expressed through individual consumption and the accumulation of private property. The population thesis as it relates to climate and environmental change cannot afford to be reduced to a quantitative problem of numbers at the expense of qualitative differences because this reduction skirts around the dynamic of exploitation in operation the world over. This dynamic enables only a few to share the outputs of labor and to do so in a largely unchecked way.

Ironies in utilitarian climate argument

4

TO BE OR NOT TO BE THIRSTY

s the earth's atmosphere warms, the climate is changing. This situation is causing extremes in the hydrologic cycle, placing the world's water resources and the species and ecosystems that depend on them under serious stress. The CO2 content of the earth's oceans is currently increasing at a rate of approximately 2 billion tons a year, causing ocean acidification. Since industrialization, the oceans' acidity has grown 30 percent, and this situation has negatively impacted the thousands of calcifying organisms on earth (mussels, coral reefs, algae, plankton), posing a serious threat to the fragile marine ecosystem.[1] Between 2002 and 2006, Greenland lost thirty-six to sixty cubic miles of its ice sheets; and between 2002 and 2005 Antarctica lost thirty-six cubic miles.[2] In 2006, severe water shortages throughout wildlife reserves and parks in Rajasthan, India, resulted in the deaths of monkeys, chinkaras, and cheetal deer. In June 2010, landslides and floods in Bangladesh killed fifty-three people. In January 2011, the Australian city of Brisbane was devastated by flooding that covered an area equal to France and Germany combined; killed twenty people; drowned many pets, farm animals, and wildlife; and ruined the habitat, food, and water sources for the wildlife that did survive. After channeling water over fifteen hundred

Colorado river

miles for the past six million years, the Colorado River is gradually dry-ing up; now it sometimes no longer reaches the ocean. Groundwater with-drawal, polluted waterways, and water projects such as dams continue to be significant contributing factors to species extinction and biodiversity loss.[3]

In August 2008, *Scientific American* featured a six-point plan to avert the global water crisis, and the plan was accompanied by the following head-line on the front page: "Running Out of Water." In April 2010, *National Geographic* ran a special issue titled "Water Our Thirsty World," which included several detailed maps of the world's shrinking freshwater reserves. This publication was followed by a series of articles in the *New York Times* from September to November 2010 that focused on India's struggle for potable water. All commentators shared the following theme: as glaciers continue to retreat and lose mass, deserts expand, rivers run dry, meat consumption increases, and the global population grows, the existential conundrum of the twenty-first century will sadly be "*to be or not to be thirsty*"!

Water covers roughly 70 percent of the earth's surface, and the total global water stock is approximately 1.4 billion cubic kilometers, with 97 percent of it being saltwater and only 2.53 percent freshwater. Put differ-ently, 3.5 million cubic kilometers of the world's water is fresh.[4] Only 0.036 percent of global freshwater supplies can be found in rivers and lakes. A large amount is locked in glaciers and ice caps (1.6 percent); a small frac-tion is in aquifers and wells (0.36 percent). The remainder resides in the bodies of animals and plants and in the air.

When solar energy is consumed, it does not negatively impact the net amount of energy produced by the sun, regardless of how much is used. For this reason, solar energy is a renewable resource. Unlike solar energy, however, water is both a renewable and a finite resource.[5] Because the earth is a closed system, just one part of which is water, the earth neither gains nor loses water (water is recycled through processes of evaporation and precipitation). When used responsibly, water supplies regenerate. However, this regeneration requires that water resources be used within renewable limits. In addition, not all water remains potable when recycled (such as when it is polluted).

The UN predicts that by 2025 two out of three people will be living in conditions of water stress, and 1.8 billion people will be living in regions of absolute water scarcity.[6] Our bodies consist of anywhere between 55 percent and 78 percent water. We need it to quench our thirst, for basic sanitation, for energy, to cook, and to grow food along with other crops (such as cotton).

water shirt storm coming

A single person needs approximately twenty to fifty liters of water per day to meet his or her basic survival needs (drinking, cooking, and cleaning).[7] Therefore, with the current world population at 7 billion people and projections for population growth to peak at 9.22 billion in 2075, it is not surprising that freshwater shortages are projected to worsen dramatically.[8] We also need to heed some caution here because the ratio of water consumption to population growth is not consistent. The World Resources Institute has reported that over the course of the twentieth century global water consumption rose more than twice the rate of population growth, increasing sixfold between 1900 to 1995.[9] The figures suggest there are other important contributing factors to the surmounting water crisis than population growth, and it is this discrepancy that I intend to unpack here.

Thirst affects existence; it impacts what a body can do. Approximately one in eight people lacks access to safe drinking water, and 3.5 million people die annually from water-related diseases.[10] If 443 million school days are lost each year because of water-related illness, and millions of women and children spend hours collecting water every day instead of attending school, caring for family, or doing income-generating work, then the costs associated with water access are high indeed.[11] Factor into this assessment the lack of access accrued over time,[12] which is important to consider because it means that if one country has consistently enjoyed 100 percent access to improved water, but another has not, then the cumulative disease and economic burdens of the country that has low rates of access will be far greater than those for the country that has consistently had a high percentage of access.

Creating systems through which all people, species, and ecosystems can access the water needed for survival will be one of the defining political issues of the twenty-first century, and models of water governance vary. Some common solutions informing water management and governance include decreasing water wastage by increasing the cost of water, instituting more state regulation of water supplies, and encouraging local communities to develop their own institutions of water governance. The focus of all of these systems is human health and well-being. The invisible party continues to be other species and ecosystems.

Changing individual patterns of consumption can improve available global supplies of potable water. The growing global middle class that can now afford to eat meat and, more important, wants to change to a high-meat diet, which I discuss in more detail in chapter 6, is by default increasing global water consumption. Between 1999 and 2001, global meat production was 229 million tons, and it is expected to grow to 465 million tons by 2050. If this prediction is correct, water resources will be placed under tremendous pressure, especially when we consider that the typical meat eater's diet requires approximately 5,400 liters of water a day to maintain.[13]

So at the risk of seeming paranoid, I would like to share with you the water footprint of a seemingly modest day of food consumption for the average person living in a high-income country. The hidden freshwater costs (depletion and quality) associated with everyday consumption is called the "water footprint." You are not consuming one cup of water with that cup of coffee you enjoy first thing in the morning, but rather 140 liters of water. Similarly, as you start your day with a bowl of Kellogg's Mini-wheats and milk, stop to eat a roast beef, cheese, and lettuce sandwich for lunch, and later sit down to a bowl of meatballs and pasta served with a generous grating of fresh parmesan cheese for dinner, followed by a slice of carrot cake with cream cheese icing and ice cream for desert, you need to be reminded that it takes 1,000 liters of water to generate one liter of milk, 1,350 liters of water to produce one kilogram of wheat, and, more disturbing, 16,000 liters of water to produce just one kilogram of beef.[14]

There is mounting research on water-intensive products such as beef and cotton that stresses the importance of considering both the amount of freshwater needed to produce a product along with the freshwater used throughout the supply chain and in particular what country's water resources are being tapped to produce a product. This consideration is the basis of Arjen Hoekstra's concept of the water footprint. It takes into account direct water use, such as the water used to water the lawn, and indirect "water used in the production and supply chains of goods and services consumed."[15] The concept is not restricted to calculating water volume. It also takes into account the complex geography of water usage—such as different kinds of water (blue water or surface and groundwater, green water or rainwater stored in soil as moisture, and gray water or water needed to absorb pollutants); where water was used (if it came from an area with an abundance of water or not); and when it was used.

One common failing of statistics on population growth and increased water consumption is that they do not consider the hidden differences in water-consumption patterns between high-, middle-, and low-income countries. It is this hidden distortion that represents the biggest challenge to the sustainable extraction, usage, supply, storage, and overall care of the world's freshwater resources. The concept of the water footprint introduces a critical structure into how we calculate global water usage because it encourages us to take into account the connection between capital, transnational politics, and power as well as, in turn, the inequitable social arrangements this system produces. It also highlights the manner in which water is a commons that traverses the borders of nation-states, ethnic identity, class, gender, race, and species. And because water is an important commons for the well-being of all life on earth, the equality between the poor and the wealthy, the private and the public, men and women, human and other than human species and ecosystems is being fought over. For instance, the scarcity of water in the Middle East has been the source of disputes among Israel, Palestine, Syria, and Lebanon, all of whom stake a claim in the waters of the River Jordan. Further, when the inequitable relations defining capitalism enter the realm of water politics, the geography of the hydrologic cycle is striated through acts of enclosure, dispossession, and appropriation. In this regard, water access is in the first instance a political problem of common life that traverses the boundaries of nation-states, regions, private property, and habitats.

The water crisis is unsurprisingly capturing the attention of social activists, journalists, and politicians, and it is being billed as a problem of far greater magnitude than the looming oil crisis. The reason is almost too obvious to state: a person might be able to live without food for several weeks, but she cannot survive without water for more than a few days. As a result, there is a lively discussion over how to avert the crisis most effectively by restructuring systems of water management. This discussion has spurred a blossoming water market that has facilitated the privatization of water infrastructure, resources, and technologies.

PRIVATE GOVERNANCE

In 1999, the Bolivian government auctioned off Cochabamba's (the country's third-largest city in the Andes Mountains) public water system SEMAPA and oddly received only one bid by the consortium Aguas del Tenari, which consisted of International Water Limited (United Kingdom), Becthel (United States), and Edison (Italy). The Bolivian government took less than $20,000 as a down payment, signing a $2.5 billion forty-year concession. Many residents subsequently had to spend up to one-fifth of their income on water as rates soared. Community wells also came under the

Privatized

water

control of Aguas del Tenari regardless of the fact that the company had not dug them. In January 2000, a coalition of community, labor, and human rights activists formed the consortium Coordinadora por la Defensa del Agua y de la Vida (Coalition in Defense of Water and Life), and they protested against the water cutoffs and rate hikes. The city was brought to a standstill for four days.

To break up the standoff, President Hugo Banzer promised the protesters that the city's water rates would be lowered. When the government failed to fulfill its promise, protesters took to the streets once again on February 4, 2000. The military was brought in to break up the march with tear gas. The government then ordered the leaders of La Coordinadora, including the outspoken Óscar Olivera, to be arrested. Despite this action, the people continued their march, and on April 8, 2000, a "state of siege" was declared. Finally, the government was forced to void the contract with International Water Limited, and the state water utility company SEMAPA was returned to the people.[16]

In response to the Bolivian water riots and after having its contract revoked, Bechtel sued the Bolivians for $50 million in damages despite investing only $1 million. In response to international outrage regarding the lawsuit, Bechtel rescinded it.[17] The economic beneficiaries of Bolivia's "two-decade dance with neoliberalism" include foreign corporations, Bolivian government officials, and other leaders who profited from burying the nation in foreign debt.[18]

The Bolivian water wars cast an international spotlight on the issue of water privatization.[19] If we were to join the dots forming the argument used in support of privatizing water governance, our line of reasoning would look like the following. First, as the global population increases, potable water supplies will be placed under more stress. Second, water supplies will be stressed because the real "value" of water is not reflected by the cost. Third, accurately pricing the cost of water will provide consumers with an incentive to use scarce water resources more efficiently and sparingly. Hence, the old model of water governance—the management of water resources and services by the government, regional officials, and local groups—needs be replaced with a new model that better reflects water as an economic value and good. According to this line of argument, the quality, reliability of service, and quantity of supplies will improve only if water governance is tied to the logic of the free market. In other words, as a valuable *commodity*, water needs to be privatized. *Privatization* is "an umbrella term that includes selling assets to a private company, tendering a water concession to a private company, or awarding management contracts to a private company," and this approach has won international support.[20]

In 1992, there was a shared consensus throughout the international arena that in order to achieve sustainable and efficient water use, the water sector needed radical restructuring. This restructuring was done using neoliberal economic incentives: deregulation, privatization, and competition. The World Bank's *Water Resources Management* publication (1993), the 1992 Rio Earth Summit, and the International Conference on Water and the Environment (Dublin) supported handing the governance of the world's water supplies over to the invisible hand of the free market.[21]

What is especially disturbing is how the World Bank has financed water-reform policies the world over as an integral part of its structural adjustment program. With the purported aim of promoting the sustainable and efficient management of a country's water resources, the World Bank's lending policies have in fact supported the mass privatization of water. As a stipulation in approximately one-third (at least 84) of the 276 water supply loans that the World Bank granted between 1990 and 2002, it required the lendees to agree to some form of privatization of water resources—a prime example of how the global environmental crisis is tweaked to become an apparatus through which neoliberal economic policies are advanced.[22] A report on this practice concluded that "privatization has been an increasingly important aspect of bank loan conditions."[23]

As the practice was described to economic hit man extraordinaire John Perkins the first day on the job with consulting firm Chas. T. Main, where he was to hired to convince underdeveloped countries to accept large loans from the U.S. Agency for International Development and the World Bank,

> We're a small exclusive club. . . . We're paid—well paid—to cheat countries around the globe out of billions of dollars. A large part of your job is to encourage world leaders to become part of a vast network that promotes U.S. commercial interests. In the end, those leaders become ensnared in a web of debt that ensures their loyalty. We can draw on them whenever we desire—to satisfy our political, economic, or military needs. In turn, these leaders bolster their political positions by bringing industrial parks, power plants, and airports to their people. Meanwhile, the owners of U.S. engineering and construction companies become very wealthy.[24]

In other words, private governance of the water commons strengthens the power of economic elites. It is this situation that has produced what Maude Barlow and Tony Clarke describe as a "global water cartel." The cartel consists of transnational corporations, the World Bank, and governments, all working in tandem over the past couple of decades to privatize water supplies. The largest water moguls include the two French companies Suez and Veolia, RWE-AG of Germany, Bechtel–United Utilities, Biwater, U.S. Water, and Anglian Water.[25]

Approximately 70 percent of global privatized water is owned by Veolia and Suez. Veolia used to be Vivendi Environment and is known by many other names: General-Des-Eaux, Onx Environmental, Dalkia, and Connex, among others. The water and wastewater units of Veolia Environment serve more than 110 million customers in eighty-four countries. In 2005, it was reportedly number 463 on the Fortune 500; its net income was $2.58 billion; its net revenue was $35.96 billion; and 40 percent of its sales came from its water services.[26] *Affermage* is the business model used by the company, and it refers to a public–private partnership or *corporate welfare*: the public pays for its water infrastructure, and Veolia Environment and its shareholders profit from the investment of public funds.

A popular argument is that "partial" privatization will diversify local water supplies as a way to hedge against risk, especially for those receiving water in regions where supplies are low or volatile. The public is assured that the public–private water-governance institution will bear none of the upfront risks and costs associated with developing the plant, only the benefits of reliable water at a price they would have paid anyway. This arrangement really translates into the public's bearing the brunt of the burden for long-term investment and the costs associated with maintenance. Meanwhile, the private sector is in charge of the less costly management of water supplies.[27]

PUBLIC GOVERNANCE

Left-wing theorists have understandably responded forcefully to the neoliberalization of the world's water, advocating the visible hand of the state and local communities govern the water commons. They insist that because water is a basic need for life, it ought not depend on one's ability to pay and should therefore remain under public control. This position is divided, however, over how best to realize the goal of public water

Models of water

governance. One camp favors a model of vertical governance, an old model whereby regional and national governments regulate, manage, and control water services and supplies (both Venezuela's president Hugo Chávez and Bolivia's president Juan Evo Morales adopted this approach). However, there is a growing sense of dissatisfaction over the complicity between the state and private sector as well as a lack of confidence in the government to maintain public water infrastructures adequately and to supply all corners of the population equitably with the water they need. This sense of dissatisfaction has recently resulted in a wave of theory in support of a horizontal approach to the governance of the commons (by Michael Hardt, Antonio Negri, and Elinor Ostrom, to name a few). This group argues that this theory is in favor of the management of local water supplies by local collectives and associational institutions. Both positions hold a great deal of validity and are enormously helpful when it comes to democratizing the management of and access to water supplies.

The horizontal approach of public water governance favors local self-organizing systems of management. Hardt and Negri can certainly be situated within this camp. In *Commonwealth*, they define two senses of the term *commons*—the first being the earth and its ecosystems and the second being the creative commons, which includes indigenous knowledge, ideas, images, and affects (service and care).[28] They recognize that when faced with a crisis, we tend to rule either in favor of more governmental regulation and control or for the neoliberal model of increased privatization. They, however, are more interested in trying to forge an alternative to the private-property system and to "institute a shared common wealth." The practical question, though, is: How do we go about *justly sharing* that which is held in common?[29] This is also a question that guides David Bollier's discussion in *A Silent Theft: The Private Plunder of Our Common Wealth*. It is this commons dilemma that guides the research agenda of political scientist and 2009 Economics Nobel Laureate Elinor Ostrom.

commons dilemma

In particular, Ostrom has extensively examined the question of whether communities are capable of effectively managing their own water resources without relying on government regulation or corporate privatization.[30] She has used the principles of game theory to study the commons dilemma.[31] What is especially innovative about her work is the tripartite structure of her method. She uses inductive theorizing in combination with field studies and lab experiments. In her collaborative work with William Blomquist, Ostrom returned to the California West Basin groundwater reserves, where she had previously studied the role of local groups managing the water basin for her 1965 doctoral dissertation.[32] Together they assessed whether the partnership between nongovernmental and governmental groups had effectively managed the water basin.[33] Their findings point to the importance of locally generated institutional arrangements that emerge out of informed open communication between common-pool resource (CPR) users.

With Roy Gardner and James Walker, Ostrom went on to develop a series of baseline experiments to examine some of the hurdles CPR users encounter when trying to achieve outcomes they share in common.[34] To clarify, CPRs are "natural or man-made resources whose yield is subtractable and the exclusion from which is nontrivial (but not necessarily impossible)."[35] Their experiments allowed subjects to earn money by either appropriating the CPR or engaging in private activities. The commons dilemma arises from the individual and social costs associated with the rational choices individuals make. It is also premised upon the notion that CPRs are both an economic good and carry an economic value. Although the payoffs from appropriation might initially be high, they decrease as other users appropriate the resource; the payoff is eventually less than that from private activities that produce a steady but marginal return. Ostrom and her colleagues found that the more individuals could communicate, the more beneficial the results were in overcoming the social dilemma. Subsequent experiments on punishment demonstrated that punishing free riders, regardless of costs, successfully lowered negative CPR appropriation rates because individuals were willing to bear the costs of punishing free riders if doing so proved to be an effective tool of deterrence. Drawing attention to a model of punishment and deterrence in this way highlights how much Ostrom frames her conception of subjectivity and political action negatively.[36]

Ostrom's observations and experiments around collective-action problems provide an intriguing look at how people work together to change their environment for the benefit of the majority. One of her work's greatest strengths is its empirical basis, for it allows her to move beyond the abstractions of theory. Through her studies of real-world problems and how individuals overcome them collaboratively, she provides us with a useful revision of vertical modes of governance that typify state, national, and international policy.

Nevertheless, the self-management model marginalizes transnational and supralocal power relations as well as the way in which these relations structure the commons and its products. For example, the CPR model does not account for the amount of dependency any given community might have on the water resources of another community, region, or country—in other words, the water embedded in the products they consume (virtual water) or how much they rely on foreign water resources in their daily products and energy needs or the water footprint imbalances this reliance produces through the importing of water-intensive products and the exporting of less water-intensive products.[37] Factors that influence the importation of water-intense products include water scarcity, lack of fertile land, and scarcity of other raw materials. A case in point: between 1997 and 2001, the virtual water flow was reportedly 1,625 billion cubic meters per year accounting for 16 percent of global water use.[38]

Ostrom's analysis and experiments focus on what motivates individual CPR users and in particular the importance of endogenous institutions in motivating the negative subject to act in the interests of the collective. The motivational focus of her work echoes current philosophical approaches to the contemporary subject. Simon Critchley points out that in response to the eternal state of war characteristic of contemporary life, people are suffering from two kinds of deficit: motivational and moral. Together these deficits have produced a lethargic subject, one who is devoid of conscience and is politically uncommitted.[39] Critchley proposes that democracy can be brought back to life if and when the subject undergoes a renewed sense of commitment to the "unfulfillable demand" posed by the other, which he explains would prompt a critical anarchic movement to emerge from the ground of the social field—a movement that is resistant to the order imposed from above. His model of collective action shares sympathies with Hardt and Negri's in that he too embraces the anarchic *potentia* of the critical ground of the social field, yet unlike in Hardt and Negri's work his subject is motivated by a lack or by a general sense of dissatisfaction and inadequacy. That said, Critchley's politically engaged subject is quite different from the fully coherent and rational subject that Ostrom takes as a given, and the negativity defining his subject is different from the negative role that motivation plays in her experiments.

Ostrom never questions the political validity of the liberal notion of the independent free-choosing subject. On the one hand, she strongly advocates for the local, a position usually articulated in alignment with the specificities of place or region. On the other hand, she invokes a universal category—the rational subject—that is anterior to spatiotemporal configurations informing local subjectivity. This tension can have considerable political consequences—namely, stripping local subjects of the very conditions that orient and locate them in a specific context is a neoliberal ideological strategy.

In addition, Ostrom does not tackle the geopolitics of water resources and services (the global dynamics of power are artificially kept separate from the local). Her appropriators of CPRs are considered in isolation to the local attachments and histories that produce and motivate her subjects. This isolation gives all the more power and credibility to the invisible hand of the free market to extend its influence. It is only by virtue of the local, as informed by the forces and energies that make up the global, and their overlapping histories that the local is allowed to articulate its specificity. Granted that this process of articulation is not easily subsumed into a model of identity politics, it is important we recognize that although specificity is distinct from the global whole, it is nonetheless informed by that whole. We do need to heed a word of caution here: although it is important to recognize local specificity, it is equally important that the local is not fetishized. One way to overcome this fetishization is to engage dialectically the differences that "make" a subject.

The local functions politically when it articulates its specificity vis-à-vis the global. In this light, focusing on the rational individual who is free to make choices can constitute an act of political erasure, for it removes from the picture the dynamic interaction of global and local forces and energies that condition subjects and complicates places. This approach also neglects to consider the sense of intimacy that informs how people relate to water. For example, our relationship to water in the shower, on the toilet, in the kitchen, in the garden, or at the river differs across individuals, communities, and cultures. These differences inform the trade-offs in water usage that people and communities make. So because the controlled environment of the lab experiment is an "ideal" experimental CPR situation, devoid of the material exigencies that inform real-life situations, it does not recognize the material forces that constitute and organize subjects, and that positions subjects differently within a local setting and the larger national and transnational community. This is how the biopolitics of water production and exchange works.

Biopolitical production points to "new mechanisms of exploitation and capital control," as Hardt and Negri recognize. In turn, the dynamic of exploitation at work in capital accumulation, they remark, has taken

labor vs. Capital
turns into

the "form of *expropriation of the common*." Their understanding of the common is both the wealth of the material world that we share in common (air, water, earth, etc.) and the "results of social production that are necessary for social interaction," such as information, affects, and knowledge. In the biopolitical era, the Marxist contradiction between the "*social* nature of capitalist production and the *private* character of capital accumulation becomes ever more extreme."[40] Hardt and Negri go on to revise Marx's constitutive political antagonism between labor and capital as one between the common and capital. I do not deny that there is a struggle occurring between the common and capital, but I am less willing to let go of the antagonism between labor and capital because I believe that if we do, we then marginalize the problem of "working out how the law of *value* operates," as Marx put it.[41] Value is an important conceptual ingredient in an analysis of how capital and environmental politics intersect.

The form of the historical circumstances in which the law of value operates today is neoliberalism—privatization, individualism, the free market, deregulation, and the ideology of the liberal subject; and the modality of that law of value is biopolitical—the production of social life, the effect of which is not the common, but the exchange-value of social life. The common is not conditioned by the biopolitical; it is necessarily antagonistic to the biopolitical, and we must keep this analytic distinction intact because without it the common is always already co-opted. It is this co-optation that so many activist groups are struggling against and that forms the basis of a variety of political movements. Part of the struggle that labor is engaged with in the age of neoliberal capitalism is over the equitable distribution of access to the commons, which is increasingly difficult the more capital monopolizes it. Hence, the definition of the commons as inherently political cannot become the same as the capitalist definition of the commons as biopolitical production. The distinction is once again machinic: the capitalist distributes access to the commons by means of the rate of profit, not by means of the political principle of equality.

For Marx, capital was always already a mode of social production, and the way to understand this definition is through the antagonistic relationship between labor and capital. So the explanatory power of the concepts of labor and capital are all the more important as an analytical distinction in understanding the biopolitical production of capital that Hardt and Negri identify as the conditions producing the common. Hardt and Negri see the value necessary for social production as rooted in the common. For them, the common can be both a revolutionary *potentia* as much as an oppressive *potestas* (biopolitical production of subjectivity in place of the old model of commodity production). I find the first part of their formulation incredibly useful but would like to revise the second part.

The commons is only *potentia*, for if we take the liberty to stretch the definition to make the commons also an oppressive force, we lose an important analytic distinction between biopolitics and the commons. Biopolitical appropriation does not change the "common"; it changes the formal relationship of nondiscriminatory reciprocity that conditions the commons as shared. And the violence over access to and control of the world's water resources is an example of the latter change.

To appreciate fully how capitalist biopolitical production works in the context of the water crisis, it is helpful to scrutinize closely the geopolitics of water governance. Only 8 percent of freshwater supplies are used for domestic use; the rest of the world's water is consumed by industry (22 percent), by farming for irrigation purposes (70 percent), and by the expanding for-profit system of supplying and managing water.[42] If we focus too much on the issue of growing human demand for water, we run the risk of resorting to a form of Malthusianism—pointing the finger of blame at the growing global population without recognizing that not everyone consumes water equally the world over and that other-than-human species remain invisible altogether in the dominant political discourse surrounding water equity.

The first approach to horizontal governance gives weight to government institutions. It advocates for a formal model of governance. The second approach favors the informality of civic life. In an era of global capital, it is also important to try and bridge the two positions, all the while recognizing that not everyone or every species can participate equally in political life.[43] Greater clarity is needed over the way in which asymmetries of power facilitate and are the effect of the institutionalization of neoliberal economic principles shaping the global water market. Left-wing theoretical discourse therefore needs to address first who the subject of struggle in the water crisis is and then the subsequent problem of how that subject can best be heard.[44] To do this, our analysis has to move beyond the empirical study of water as an isolated substance or hydrologic cycle or both.

WATER PLUS CAPITAL

Let us begin by breaking down the global average water footprint of 1,243 cubic meters per capita per year. The per capita annual water footprint of

China is 702 cubic meters, with only 7 percent of Chinese water consumption coming from offshore. The water footprint of somebody living in the United States is 2,483 cubic meters per year, with 19 percent of that coming from outside the country. Meanwhile, in Kenya the average per capita water footprint per year is just 714 cubic meters, with 10 percent coming from outside the country.[45] There is a clear correlation forming here between the inequitable distribution and use of water and the inequitable structural dynamics of the global economy. And the geography of the water commons does not stop with a comparison of per capita footprints. It also investigates and interrogates the growing market in water investment, production, and infrastructure.

As the material flows of the hydrologic cycle—a process of evaporation (from bodies of water such as lakes), condensation (clouds), circulation (through atmosphere), precipitation (rain and snow), and dissipation (run off)—are distributed and differentiated by capital, that cycle is placed in the service of capital accumulation. For example, water is distributed by wells through processes of desalination, sewerage systems, or dams. It is differentiated as a private investment opportunity (as an economic good and holding an economic value), a public utility (wastewater-treatment facilities, public drinking-water facilities), or a commons (rivers, lakes, rain, and oceans).

As water is distributed and differentiated, it also enters a process of signification—for instance, when it is defined as a water market. And this process in turn organizes social subjects—namely, those who can afford to pay for water services and those who cannot. This organization concomitantly defines who counts as a social subject and who does not. In other words, the needs of the poor living in urban slums who do not have access to formal water infrastructure or the wildlife that are forced out of their habitats when water sources run dry do not count, for how they consume water circumvents formal water services and the surplus-value generated from these services.[46]

The continual movement of water over, under, and across the earth is a form of social production and reproduction. Unlike capital, however, the hydrologic cycle is a closed system. As such, it does not produce a surplus. Contrary to the logic of capitalism, the hydrologic cycle is healthiest when it operates as a closed loop. So how do capital and the hydrologic cycle connect if capitalism is, as Marx posited, a process of capital accumulation and in this instance what is being accumulated is a closed circuit?

First, the value produced by the hydrologic cycle has to move beyond utility (use-value) and the preservation of value. It is reconfigured to generate an exchange-value, whereby additional value is created and a surplus is produced.

To account for the way the production process of the hydrologic cycle is organized to make a profit, it might be helpful to turn to Marx's labor theory of value. In *Capital*, Marx explains that capital accumulates by exploiting labor. To put it in starkly simplistic terms, the laborer sells his or her labor power to the capitalist in return for a living wage as the capitalist generates a surplus-value from that labor. For Marx, the important point in all this is that capital accumulation depends on two important transformations taking place. First, labor power has to be turned into an exchange-value. Second, as labor is exploited by capital, it is alienated. However, Marx did not regard water as having a value for the simple reason that it is not produced with human labor but rather is part of the life cycle. As material life has been neoliberalized, this situation has dramatically changed. If we were to put the notion of labor-value to work in a somewhat idiosyncratic way, we might consider the productive forces of the hydrologic cycle that were previously independent of capital as now subsumed by capital. By transforming the movement of the hydrologic cycle into an exchange-value, the effects of the labor of material life are brought under the control of capital, which alienates material life from its ecological cycles (precipitation and so on) and its creative capacity (reproducing life).

For instance, one popular new technology that allows people to tap into new supply options and increase the productivity of the hydrologic cycle is desalination. The desalination process removes salt and other minerals from seawater to make it ready for industrial use or human consumption or both. The popularity of desalination plants, especially for governments, is that they can provide a reliable water supply with high water quality. Nevertheless, there are many unknowns associated with this technology. Desalination can also introduce chemical and biological contaminants into the water supply; produce water that is corrosive to water-distribution systems; create high concentrations of salt brines, leading to the problem of how to dispose of the effluent safely; negatively impact marine organisms; and lead to greater dependence on fossil fuels, which in turn produce GHGs and further the warming of the climate. Then there is the pink elephant in the room: the private ownership and operation of desalination facilities.[47]

History of desalinization

And although there is legitimate cause for concern over the environmental impact of developing and running large-scale commercial desalination plants, I am equally concerned over the use of public funds to support the research and development of desalination technologies, which are eventually appropriated by the private sector under the guise of providing a public service.[48] A brief trot through history helps make my point.

The idea of separating salt from seawater is not new. In 1790, the U.S. secretary of state Thomas Jefferson was asked to sell the government a distillation system that could convert salt water to freshwater. In 1852, a British patent was given for a distillation method that did just this. In 1952, the U.S. Congress passed the Saline Water Conversion Act (Pub. L. 92-60), which would fund the Office of Saline Water in the Department of the Interior's Bureau of Reclamation. President John F. Kennedy was reportedly a strong supporter of large-scale commercial desalination. In 1977, the United States spent approximately $144 million on desalination research, and further public funds were allocated to offshore desalination efforts in other places such as Japan and the Persian Gulf. After a brief lull during the Reagan administration when public-research funding went to military initiatives, in 1996 Senator Paul Simon (D–Ill.) brought into effect the Water Desalination Act (Pub. L. 104-298). The act provided $30 million in federal support over a six-year period toward desalination research, along with a further $25 million between 1999 and 2002 toward demonstration projects. Interestingly, the fiscal year 2005 Omnibus Bill modified this legislation, requiring the private sector share 50 percent of the costs associated with these ventures, and yet the U.S. government appropriated only $2.5 million for the 1999 fiscal year and $1.3 million for the 2000 fiscal year.[49]

In 2009, Tom Pankratz of the International Desalination Association estimated that the global seawater desalination market might eventually reach $58 billion dollars over the next ten years. Those at the forefront of the burgeoning market in desalination include the usual list of multinational water moguls: Veolia Environment (France), which is desalinating 5.4 million cubic meters a day; Fisia Italimpianti (Italy), 3 million cubic meters a day; Doosan (Korea), 2.8 million cubic meters a day; and GE Water (United States), 2.5 million cubic meters a day.[50] In 2009, chief marketing officer for GE Water, Jeff Fulgham, expected "double digit growth rates in the [desalination] market." As he went on to admit, "That is a big deal for us from an investment standpoint."[51] Not bad, considering his prediction was made during the height of the global economic meltdown.

TRANSVERSAL GOVERNANCE

Social relations (intersections of class, race, gender, and sexuality), material life, technology, reproduction, mental frameworks, and modes of governance and organization are distinct but at the same time dialectically implicated in each other. The horizontal governance model of the commons needs to be fine-tuned with a vertical system of governance that recognizes the borders framing local, regional, and national landscapes as well as the transnational flows of power and privilege that define the geopolitical arena. We need to cast a critical eye over the institutionalized systems of oppression and hierarchy that obscure the visibility of some individuals, communities, species, and ecological systems from consideration. That is, we need to be mindful of how the intersection of labor, power, and capital impedes the visibility and ethical considerability of marginalized groups: the poor, women, other-than-human species, and ecosystems.

Neither privatization nor vertical or horizontal governance models adequately engage with the asymmetries of power shaping the social field. Social arrangements, regardless of operating at a more intimate and "local" scale, do not work equally for everyone.[52] For instance, although Cochabamba's water was returned to the people after the protracted water wars, longstanding class differentiations and a culture of political corruption meant poor residents still struggled for reliable and safe access to water. In addition to issues of class, there are also endemic gender biases that inform the water debate the world over. Water is not gender neutral, especially in low- and middle-income countries. Accessing water for subsistence agriculture, basic health and sanitation needs, and domestic consumption needs is primarily the role of poor women in these regions.

In many parts of the world, water collection is women's work. When women are not included in the management of water projects or programs, their water rights and privileges are often not met. When water supplies are scarce, it is women who spend more time in the day traveling to water sources and girls who are removed from school to assist in the collection of water, so that less time can be spent on subsistence agriculture, simply reinforcing the acute asymmetries of poverty that disadvantage women. A case in point is the Macina Wells project in Mali. It failed because the notion of a community-managed well was blind to differences in gender. Management of the wells was allocated to the men, and yet the women were the ones responsible for collecting water. The failure to consult women in both the planning and the management of the well resulted in equipment the women found impractical to use, so that they eventually removed it; moreover, the men failed in their management duties because they regarded water and sanitation as women's business.[53]

Institutionalized gender inequalities present some serious obstacles for the model of horizontal governance for the simple reason that not all social arrangements allow for the equal participation of all members.[54] In the planning of water programs, skewed gender relations are not sufficiently addressed by inviting women to participate. For example, when a water project was being planned for the Tanga Region in Tanzania, women were not present for three reasons. First, the meetings were held at a time that was impractical for them to attend. Second, they felt the men would not seriously listen to their suggestions. Third, they were not properly informed about the meetings.[55] Further, women's participation in the management of local water supplies is not an automatic panacea for equality. How can the inequities of unpaid labor be addressed? Will childcare facilities be available to allow women to spend the time needed for managing a given program?

The disquieting transformation of freshwater into a liquid asset is also a story of the intrusion of market criteria in life. Moreover, the lucrative speculative economy that is growing around the predicted water crisis is driving the push to privatize the water commons. One way to combat the privatization of the commons is for water resources to remain a public utility, and in countries where this is not currently the case, the solution would be to nationalize water resources and infrastructure.[56] This is exactly what Bolivian president Evo Morales did once he was elected to power in 2006.[57] Yet since 1970 glaciers in the Andes have lost 20 percent of their volume, and because Bolivia draws 30 percent of its water supply from the glacial basin, the water situation there is becoming critical. The disappearing glaciers have resulted in severe water shortages in parts of Bolivia, especially El Alto, where domestic water taps have reportedly dried up. The World Bank reports that nearly 100 million people might be adversely affected by the melting glaciers.[58]

In other words, stringent national government regulation of water cannot solve the critical water shortages incurred because of climate change. The solution requires a multilateral approach where broader alliances are formed throughout the international community to slow the warming of the earth's climate and to help the most vulnerable, including other-than-human species and ecosystems, cope with dwindling water supplies. Consider the case of Bolivia: its water shortages are in large part the creation of the market—the long history of industrialization and capitalism that has produced changes in climate, whose impacts exceed the borders of nation-states and which are primarily responsible for the GHG emissions buildup causing ice caps to retreat.

A transversal mode of water governance, one that is mobilized across different geographic scales and that combines horizontal and vertical governance models, is urgently needed to address the transnational character of our water commons. We also need to be mindful of how different power relations create the water crisis and of how this crisis is then used to facilitate a neoliberal agenda that has gradually realized the privatization of the water commons. In light of the corrupted agenda of the World Bank's efforts to administer the governance of local water, far more transparency needs to be enforced around the conditions of supposed "aid" and "relief" loans and the various corporate interests such loans support, along with more stringent measures of accountability for how international aid organizations do business.

Although individual interests and choices as well as equitable modes of governing the water commons are important, equitable access to freshwater involves a larger problem of ecological organization. What is needed is a dramatic change in how we relate to one another, our environment, other species, our pasts and futures—all with a view to forming alliances across generations, species, and communities. How does the water crisis generate solidarity among individuals and across communities? This question is not a contractual or legal problem; it concerns the particular attachments people have to a place, the histories they share in common, and how these histories shape and are shaped by the places in which they live. How can we arouse a shared feeling for equality that traverses species? How can we move from "limited sympathy" to "extended generosity"?[59] Moreover, the question of how solidarities are affected and in turn affect the places in which they live might elucidate future political trajectories.

The social and environmental injustices that the water crisis compounds cannot be fully accounted for by a political theory that focuses exclusively on choosing between vertical or horizontal modes of governance. It needs to shift to a temporal organization that commences with the proposition that if we continue on our current course, by 2050 water scarcity will hinder the flourishing of all life on earth. The question is: What can we do today to stop this process? We also need to understand the water crisis not merely as an effect of the failure of individuals and groups to develop formal and informal rules that effectively manage the resources they share in common. Rather, we need to capture how the unequal distribution and exhaustion of the world's water resources is rooted in the asymmetrical power relations and social orders emerging out of capital flows and to recognize that this relationship is connected to the neoliberalization of life.

The water commons dilemma marks a breakdown at the level of social cooperation. I am using the term *social* in its expanded sense to include human beings, other species, environments, ecological systems, and future generations. The dilemma presents a fundamental failure at being inclusive. Now more than ever we cannot afford to forget the ominous pronouncement of French underwater eco-explorer Jacques Cousteau: "We forget that the water cycle and the life cycle are one." This leaves us with the following question: How might commons dilemmas, such as water access, engage a critique of capital? For that matter, how can commons dilemmas generate alternatives to the neoliberalization of life?

Water is a common good. Unlike for oil resources, another source cannot be substituted for water. Because water is the basis for all life on earth, access cannot be contingent upon the ability to pay. What might seem like a ludicrous example brings the issue into focus. The water company is going to be hard pressed to figure out how it will invoice the local bird population for drinking from its water supply! More important, though, the birds are going to be hard pressed to access the water they need when surface-water sources run dry or are enclosed through privatization.

5

SOUNDING THE ALARM ON HUNGER

You're either a hustler or you're bein' hustled.

—Professional gambler Billy Walters[1]

A longside water, food is a fundamental human need. As of 2010, the number of undernourished people in the world was nearing the 1 billion mark, more than the estimated number of hungry people during the 2008–2009 food and economic crises.[2] If population numbers increase to more than 9 billion people as projected, and the demand for animal protein continues to grow along with the expanding middle class in China and India, then the global food system will be placed under tremendous stress. When we also factor into the picture growing water scarcity, rising temperatures, and very few sources of new farmland becoming available, the food system may very well enter crisis mode.

Adapting to climate change as it relates to the food system is viewed primarily through the lens of technological innovation and a political commitment to extending a rights-based framework into the realm of food-security policy and program development, all the while maintaining a neoliberal economic paradigm. The larger problems arising out of the liberalization of the agricultural sector that has tied food production to free-market forces, which in turn influences price hikes, tend to remain largely in the shadows of discussions of climate change and food scarcity.

Furthermore, the concern is that as a "crisis" mode sets in, the food-production system will be further restructured to facilitate capital flows, and the deeper problems arising out of neoliberal economic policies in the global food system will remain unchallenged. New systems such as conservation agriculture might provide people with useful new strategies to deal with the havoc that climate change will wreak on the practice of farming, but remaining largely absent are the overriding questions of who owns the patents on the new climate resilient seed varieties and new seeding technologies as well as of the problems surrounding crippling farmer debt, declining biodiversity, and the monopoly that the economically powerful continue to hold over global prices of food staples and the system of food exchange.

My intention is not to undermine the important work being done on how climate change will impact the global food system and the strategies being developed to deal with this impact or the recommendations of rights-based approaches to ensuring that all people have access to adequate nutritious food. Rather, I intend to scratch beneath the surface of these debates to bring into focus other structural causes behind the global food crisis, for it is important that a difficult situation is not made worse. As we move forward to deal with the interrelated phenomena of environmental change and the need to transform the fossil fuel economy, we need to be mindful of the inequities that might unnecessarily be amplified as a result of a blinkered approach to these problems. As scientists, policymakers, farmers, and businesspeople set out to address the Molotov cocktail of climate change, water scarcity, the biofuel industry, growing demand for animal protein, lack of new agricultural land, weakening biodiversity, and more people on earth, there is a darker shadow that looms in our midst, and that is how the production and distribution of food are driven by free-market principles.

It is a basic law of nature that food production is sensitive to climate, and one of the nastier effects of climate change will be the extent to which it will disrupt the global food system. Climate change will dramatically alter current water availability, increase the frequency and intensity of weather extremes, result in more acute precipitation, and lead to sea-level rises. As such, it will seriously impact how much food is produced, the quality of that food, and the availability of different food varieties. And although human beings have to a large degree developed the capacity to control the impact climate has on food production by building greenhouses, developing irrigation systems, using cold-storage mechanisms, introducing pesticides, establishing temperature-controlled houses for livestock, and so on, these adaptation techniques will be insufficient to deal with the breadth and depth of future environmental changes associated with increased CO_2 levels that scientists predict will accompany a 2°C (or more) rise in average global temperature. As Mark Lynas so vividly describes, just a 1°C rise will be enough to revert the food-productive areas of the High Plains states in the United States to sand![3]

What can we expect? If the 2003 European heat wave and the subsequent 30 percent drop in agricultural yield Europe encountered is any indication, we are in for a rocky ride as the world heats up. Climate change will lead to greater weather variability and more extreme weather events, which will in turn lead to a decline in crop yields. For instance, what might currently be considered a very wet summer in the Asian monsoon region may be five times more frequent in the second half of the twenty-first century.[4] During periods of fruit development and germination, changing rainfall patterns and higher temperatures will impede plant growth. The hotter temperatures predicted for tropical countries may completely wipe out entire crops. One study has found that temperature increases on rice production in farmer-managed rice fields in tropical and subtropical regions across the world will be upset.[5] Given that Asia produces 90 percent of the world's rice, these findings are alarming indeed. Although it was once believed that increased CO_2 levels in the atmosphere might increase the fertilization effect of C3 crops (rice, wheat, and soybeans) by speeding up the process of photosynthesis, recent studies show that the fertilization picture for C3 and C4 (maize and sorghum) crops has been far too optimistic, with only half of what was originally predicted for C3 crops being fertilized and minimal or no effect on the fertilization rates of C4 crops.[6] Crop losses from pest infestation will increase as insect breeding seasons and rates of reproduction grow in warmer temperatures. And rising ozone levels will reduce crop yields further.

Increasingly warmer temperatures as a result of changes in climate will in turn increase the incidence of extreme climatic events, such as drought across the world. By 2025, two-thirds of Africa's arable land is expected to be lost to drought.[7] Although droughts are a standard feature of climate variation, computer climate models developed by scientist Aiguo Dai from the National Center for Atmospheric Research show that by the end of the twenty-first century the world's land area will be significantly drier (less surface water and drier soils) than in the past. Dai reports that from

1950 to 2008 most land area warmed between 1° and 3°C, with the majority of warming occurring over northern Asia and northern North America. The future for agriculture looks grim indeed, with global aridity and drought areas significantly growing. His models project that the Western Hemisphere as well as Africa, Australia, and Eurasia will need to deal with extreme drought conditions and that the western two-thirds of the United States will become significantly drier by 2030. He is careful to point out that there is a window of variability within his findings because they are also contingent upon how many GHGs are emitted in the future and upon other climate cycles (variations in climate caused by El Niño or La Niña, for instance).[8]

In the past, the effects of drought have been devastating. In the 1980s, it killed more than half a million people in African nations and caused $40 billion in damages in the United States.[9] Based on Dai's projections and others, not only will the future be thirsty, but growing seasons will be permanently altered, ecosystems will be undermined, negatively impacting biodiversity, and social conflict will arise as the reliability of food supplies is compromised.[10]

The poor will once again bear the brunt of the food-scarcity burden. Hotter weather will cause food to deteriorate more rapidly, and although there have been tremendous advances in food storage since the onset of industrialization, poorer communities and households that cannot afford refrigeration and other storage facilities will be at greater risk of already low and scarce food supplies being spent. Agriculture currently provides 36 percent of the global workforce with a livelihood, but in poorer regions of the world this figure is expected to increase dramatically. For instance, 40 to 50 percent of people in Asia and the Pacific make their primary living from agriculture, and two-thirds of those living in sub-Saharan Africa depend on agriculture for their livelihoods.[11] Along with changing weather patterns, there will be increased changes in seasonality, which will place under stress rural farmers who depend on rain-fed and subsistence agriculture; these people also do not have the income to purchase imported food. Rain-fed crop yields in Africa may be reduced by as much as 50 percent by 2020.[12] Further, the traditional methods and knowledge rural farmers use to predict climate will become obsolete, placing these communities under even more strain. Farmers who depend on a single annual harvest are especially vulnerable to such seasonal changes, as are those who rely on seasonal foods to meet their basic survival needs. The World Health

Organization (WHO) reports that "the traditional diet of circumpolar residents is likely to be impacted by melting snow and ice, affecting animal distributions and accessibility for hunting."[13]

All in all, climate change is projected to compromise global food security seriously. The World Food Summit definition of food security, formulated in November 1996, states: "Food security exists when all people at all times have physical or economic access to sufficient and nutritious food to meet their dietary needs and food preferences for an active and healthy life."[14] The four dimensions of food security are: (1) food availability, (2) food accessibility, (3) food utilization, and (4) food system stability. The term *food system* refers to the interconnected processes of food production, processing, distribution, consumption, and waste. The UN Food and Agriculture Organization (FAO) explains that a "food system comprises multiple food chains operating at the global, national, and local levels," adding: "Some of these chains are very short and not very complex, while others circle the globe in an intricate web of interconnecting processes and links."[15]

The FAO has commissioned studies that examine how climate change will impact the global food system. It predicts that the amount of food and type of food will be affected by changes in climate, as will incomes tied to the effective functioning of the global food system. It also stresses that in addition to impacting the livelihoods of already vulnerable groups, the mounting cost of energy and the need to lower fossil fuel consumption may force local communities to take on more responsibility for their own food security. However, climate change is not going to be geographically neutral: wet regions are expected to be wetter, dry regions will become dryer, and temperate regions will fare better overall. Over the past fifty years, the average global temperature has increased twice as much as it did during the first half of the twentieth century, but these increases have not been regionally the same, with changes in the temperate climates of the Northern Hemisphere and Southern Hemisphere being greater.[16]

WHO has drawn attention to the myriad ways in which climate change and the connected problem of global food security will impact human health. It has also highlighted the particularly vulnerable position of the world's poor, a population that is already suffering from malnutrition and hunger. Approximately 3.5 million people, mostly children, die from malnutrition and related diseases every year. This situation will most certainly be exacerbated if crop yields fail and already tenuous food supplies drop, leading to further hunger, malnutrition, and disease.[17] In addition to decreasing crop yields from climate change, subsistence farmers and the poor who cannot afford food are especially at risk of hunger and malnutrition, which in turn increases their susceptibility to viruses and bacteria that spread in contaminated food and water. Malnutrition weakens the immune system, making a person more prone to getting viruses that cause illnesses such as diarrhea, which further weakens the body. But it should also be noted that not just low-income countries are at risk of food-borne diseases; WHO reports that climate change is also expected to "increase rates of Salmonella and other food-borne infections in Europe and North America."[18]

How are governments and international agencies responding to these alarming predictions, and what solutions are being proposed and tested? The FAO insists that in order to meet the increased demand for food by a growing population and to offset the negative impact that changes in climate will have on food production, production should achieve a "higher yield per unit of input." For the FAO, higher yield per unit of output means improving land-management practices, maintaining and improving plant and animal genetic resources, more efficiently managing livestock and fishery production, and developing better water-storage systems for agricultural use.[19] Moreover, the FAO favors conservation agriculture, arguing that it not only has the potential to transform the agricultural sector into a more "sustainable" system but can also help lower the GHG emissions associated with current farming practices.

Conservation agriculture begun in Argentina and Brazil aims to protect and improve land resources while also using modern technologies to increase production. Conservation agriculture is not "based on maximizing yields while exploiting the soil and agro-ecosystem resources"; rather, it is "based on optimizing yields and profits to achieve a balance of agricultural, economic, and environmental benefits."[20] Some of the primary features of conservation agriculture are zero tillage, the use of laser levelers (cuts water use), crop residue mulch (increases water holding in soil), dry seeding (uses less water), drill seeding (applies herbicides, fertilizers, and seeds together), green manure (reduces water loss through evaporation), and crop diversification.

Farmers who are employing direct seeding in Haryana's Karnal region in India are reportedly using 20 percent less water. New seed drills are more precise and have smaller furrows, which helps stop seeds from drying out when the direct seed method is used. New seeders also have the capacity to plow through previous crop residues, and because fields are left untilled, this ability produces financial savings for farmers, who spend less on labor. Another conservation agricultural method is to plant sesbania (a legume) in with the rice. This combination suppresses weed growth, and Indian farmers who use the technique have reportedly saved 1,500 rupees on the expense of one hand weeding. When sesbania is killed off around thirty days later, it turns to mulch, providing approximately fifteen kilograms of nitrogen per hectare and reducing water losses through evaporation.[21]

Genetic engineering studies and trials are working hard to develop crops that have a stronger resistance to flood, drought, and salinity levels. More resistant rice varieties are currently under trial by the Stress Tolerant Rice for Africa and South Asia (STRASA) project. Phase 1 of the project focused primarily on crop breeding. It resulted in the production of 3,400 tons of seed of both popular and stress-tolerant rice varieties and the screening of more than 600 germplasm accessions for their tolerance to drought, salinity, iron toxicity, and changes in temperature, which together resulted in the selection of stress-tolerant rice varieties. In May 2011, AfricaRice, in collaboration with STRASA, launched Phase 2 of the project in Contonou Benin. The aims of Phase 2 are to develop Phase 1 initiatives further and to improve crop-management strategies.[22]

In West Bengal, where losses to rice production due to environmental stresses are high, genetically modified stress-tolerant varieties of rice were introduced to farmers. Out of 5,780,000 hectares used for rice production, 1.03 million hectares are susceptible to flooding, and 1.46 million hectares are prone to drought, with 440,000 hectares suffering from salinity.[23] The new rice seed variety is quickly moving throughout the area, with more and more farmers using it with success.

In tandem with asking how new technologies and knowledge might enable the agricultural sector to adapt and even mitigate environmental changes as a result of land degradation, changes in climate, and water scarcity, the equally important question that needs to be posed is, Who will own the patent on the new seed varieties? Who is responsible for producing and distributing the new Swarna-Sub 1 (submergence tolerant) and IR72046 (salt-tolerant) rice varieties? Will farmers be going into debt to pay for the new machinery? And will the financial returns from conservation agriculture be enough to meet debt repayments? What happens if one farmer chooses not to use the genetically modified seed? Will he be penalized if winds blow across his fields and old seed varieties "contaminate" the new seed varieties? Moreover, what impact do new genetically modified varieties have on biodiversity?

Ecologist Debal Deb, chair of the Center for Interdisciplinary Studies in Kolkata, India, is the founder of Vrihi, a nongovernmental seed bank. He described to me the seed conservation work he has been doing on his small farm in Bengal for the past few decades. In an effort to save the few remaining folk varieties of rice in India, he keeps approximately seven hundred varieties of rice in production by growing and donating the seeds to farmers. The seeds are distributed through an informal exchange network among farmers. He explained that in order for rice seeds to germinate, they need to be cultivated within two years, so when farmers switch to modern varieties and chemical farming, both of which are the basis of conservation agriculture, along with other forms of Big Agriculture, the folk varieties simply die out from disuse.[24]

Modern varieties of rice tend to be vulnerable to extreme environmental and climate conditions and changes. In contrast, folk varieties have adapted over time to survive weather and environmental extremes and are far more resilient. Deb described folk varieties in his seed bank that are flood resistant, salinity resistant, and drought resistant. When I asked him if he had spoken with officials in the Indian government about this wealth of indigenous knowledge, he gently smiled and said he had but added that the government is basically not interested. The reason why: there is no "market" for folk varieties; they are free, and as such the government cannot see the financial benefits from distributing rice varieties that do not contribute to India's GDP. In light of this response, it is important to put STRASA's work into a broader political context. The program claims that its "new" rice varieties can solve world hunger, but the gesture is nothing more than a cynical exercise in corporate promotion. There is neither nothing "new" nor nothing "innovative" about the rice varieties STRASA is trialing; the modern varieties it promotes are "copies" of folk varieties that exist free for all. The STRASA work is outright theft, pure and simple.

Raj Patel and Vandana Shiva amply demonstrate the negative effects of new technologies that bolster the power of the economic elites at the expense of local economies, ecosystems, and household economies dependent on farming for their livelihoods. Patel is clear that the problem does not lie with new technologies being used throughout the agricultural sector. Instead, the "problem is one of power and control." He explains that when the biotechnology corporation Monsanto "created a craze for its GM [genetically modified] cotton seeds," this craze "not only led to farmers becoming 'deskilled'" but also to the "collapse of an entire farming system."[25] Dr. Deb also told me that Monsanto has paid villagers to bully and intimidate him and his mother as a result of the work he does in folk seed conservation, and he has been harassed for years now by local government officials, who claim he is a terrorist insofar as they define a terrorist as a person who goes against the interests of the state.[26] Solving hunger is not in the best interests of the state. This issue is a prototypical example of institutionalized politics, corporatism, and militarism intersecting. In a similar vein, Vandana Shiva has warned of the debilitating effects of what she calls "biopiracy," the patent on life that homogenizes the dynamic creativity of life and the sociality of knowledge production and dissemination.[27] So what can be done about this situation?

Organizations such as the UN and WHO adopt a rights-based approach to deal with the intertwined problem of food security and institutionalized inequities that make the problem of hunger and malnutrition more acute. WHO adopts this approach to solving the climate change, health, food, and water combination, underscoring that all people have a right to health, which by extension includes a right to water, food, and shelter. Furthermore, it recognizes the interdependency of all these rights—that no one right can be given priority over another.[28] The basic fundamentals of a rights-based approach to climate change adaptation begins with the premise that individual persons own certain entitlements within the social, economic, political, and cultural spheres. When rights are accorded an individual, they are expected to provide that person with basic security. The rights-based approach is at its core a political project that sets out to introduce new policies that can address the imbalances arising from poverty and socially exclusionary institutions and cultural norms, with a view to creating individual entitlements. It recognizes not only that vulnerable groups are marginalized at the national or local level, but that international policies and practices can also intensify preexisting forms of social exclusion (such as gender inequities). The limitations of a rights-based approach stem from the emphasis it gives to individual entitlements. The problem of starvation and food scarcity is a collective issue. The structures and institutions that stitch the global food system together and the asymmetries of power coordinating these structures therefore need to be more closely scrutinized.

First, climate change is projected to render the basic problem of supply and demand more acute. As crop yields fail, the amount of food available to feed the world's growing population will drop. The Fundación Ecológica Universal (Universal Ecological Fund) succinctly summarizes the situation:

> Under the current distribution patterns, global food production would not be enough to fully meet the food requirements of 7.8 billion people estimated to inhabit the world in the next decade—about 900 million additional people.
>
> By 2020, when considering the impacts of climate change and population growth, global wheat production will experience a 14 percent deficit between production and demand; global rice production an 11 percent deficit; and a 9 percent deficit in maize (corn) production. Soybean is the only crop showing an increase in global production, with an estimated 5 percent surplus.[29]

In addition, because it takes three kilograms of wheat to produce just one kilogram of meat, and 33 percent of limited cropland is used for beef production, the livestock industry and the growing demand for animal protein from a rising middle class in middle-income countries are also responsible for placing grain stocks under stress.[30]

During March 2007 and March 2008, global food prices increased approximately 43 percent. Over the same period, the price of staple crops such as wheat and soybean increased 146 percent and 71 percent, respectively. Food-insecure populations currently spend anywhere between 50 and 60 percent of their incomes on food.[31] The reasons for the increase in food prices are economic (rising energy costs as well as the increased demand for food, oil, and energy by consumers in emerging markets such as China and India), environmental (poor weather conditions thwarting crop yields), and developmental (the growth in biofuel production causing price of corn to spike). And how these factors connect are driven by neoliberal economics.

Oxfam has reported that the 2007–2008 food crisis, which was caused by sharp spikes in food prices, left an extra 150 million people hungry, and it projects that by 2030 food prices will double. Oxfam has made a plea to scale up and improve the management of food reserves, which since 1990 have basically been neglected. For instance, the food-price crisis could have been offset by means of a global grain reserve of 105 million tons, with the cost of maintaining such a reserve estimated at $1.5 billion, a cost that Oxfam puts neatly into perspective: "$1.5 billion or $10 for each of the extra 150 million people who joined the ranks of the hungry as a direct result of the last food price surge."[32] Basically, what Oxfam is proposing is the development of a much needed social food safety net to protect poor populations against upsurges in food prices.

Many are pointing the finger of blame for food scarcity and the concomitant problem of rising food prices at biofuel production, and rightly so. In the effort to become more energy independent and to shift to cleaner energy solutions, the switch from fossil fuel–based oil to biofuels such as biodiesel and ethanol is a popular alternative for some. In contrast to traditional petrofuels, biofuels do not contribute to climate change and release less particulate pollution. Although biofuel production might mitigate GHG emissions by providing clean energy, it has had the unfortunate consequence of diverting food grain–growing land to the production of crops used in biofuel.[33] In his briefing before the U.S. Senate on June 13, 2007, environmental analyst Lester Brown explained that the U.S. corn crop is a crucial ingredient in the global food economy, accounting for "40 percent of the global harvest" and providing approximately "70 percent of the world's corn imports." As the United States strives to solve its dependence on foreign oil by developing 35 billion gallons of alternative fuels that come from ethanol and coal by 2017, he warned that it might be cutting its nose off to spite its face. With the fuel value of grain exceeding its food value, Brown warns that "the stage is now set for direct competition for grain between the 800 million people who own automobiles, and the world's 2 billion poorest people."[34]

Some critical realism needs to enter the policy equation here, however. Even if all the U.S. grain crop were turned into ethanol, it would fuel only 16 percent of U.S. automobiles, and the grain needed to fill a twenty-five-gallon tank with ethanol is enough to feed one person for a year, so the math for biofuel production does not compute.[35] Furthermore, according to a study completed by David Pimentel and Tad Patzek, the production of biofuels uses more fossil fuel energy than can be compensated for by the use of biofuels.[36] The math adds up only when placed in the context of neoliberal economic policy that pushes food into the muddy waters of investment, derivatives, speculation, and futures trading.

Data produced by the Earth Policy Institute show that the percentage of the U.S. corn crop used for ethanol has risen sharply since 2005. In 2005,

10.6 percent of the 2004 corn crop yield was used for ethanol; by 2009, this figure had jumped to 26 percent of the 2008 U.S. corn crop yield of 410 million tons.[37] In 2010, 35 million acres of U.S. farmland were dedicated to crops for ethanol production, heavily subsidized by the U.S. government at $6 billion a year. It therefore comes as no surprise to hear that the United States increased its ethanol production by 21 percent in 2010. So when bad weather hit the U.S. Corn Belt states the same year, the supply of corn fell sharply, and the price of corn responded accordingly. These figures seem to suggest a structural cause to the price hikes that can be neatly situated within the cycles of supply and demand: with supply going down and demand increasing, prices escalate.

Yet the Mexican corn crisis in 2006–2007 exceeds the structural analysis of supply and demand, and it helps shed light on the different threads that tie neoliberal economic policy together: the dynamics of supply and demand, speculation, food scarcity, and the broader threat to the biodiversity upon which all life on earth depends. When international corn prices shot up in 2006, Mexico was thrown into a food crisis as the price of tortillas, a staple in the Mexican diet, tripled and in some parts of the country quadrupled.[38] Many commentators blamed the inflated prices on the diversion of corn supplies in the United States to ethanol production.[39] A rise in demand for corn was to blame in part for the rise in corn prices, yet it is also too easy simply to fault ethanol for the price hikes. We must also be aware that neoliberal policies have gradually placed Mexican corn demand under the thumb of the free market.

The free-trade policies of the North American Free Trade Agreement (NAFTA) signed by the governments of Canada, Mexico, and the United States in 1992 (it came into effect in 1994) have skewed the Mexican economy toward corn imports from the U.S. corn market, slowly crippling the country's corn self-sufficiency in the process. Prior to NAFTA, Mexico imported only 10 percent of its national corn needs; by 2007, it was importing 50 percent of its corn from the United States. In 2003, Mexican trade statistics show that U.S. corn exports to Mexico totaled 8.4 million metric tons.[40] In addition, the Compañía Nacional de Subsistencias Populare, a Mexican state organization responsible for regulating Mexico's corn market and ensuring that local producers get optimal prices for their corn, was liquidated in 1999 as a result of neoliberal restructuring measures inflicted on Mexico after the debt crisis of 1982, which further weakened the Mexican corn economy. From then on, the Mexican corn economy was slowly handed over to multinational grain corporations. I say "slowly" because the Mexicans, realizing the negative social and economic impact that receiving U.S. corn exports was having on the livelihood of Mexican farmers, bargained to extend by fourteen years the transition period from a Mexican corn market in Mexico to a market dominated by U.S. corn (from January 1, 1994, to January 1, 2008).

In what would have to be a prime example of corporate opportunism at its best and most ruthless, Monsanto rubbed salt into open wounds when it used the corn price crisis to push for more genetically modified corn crops to be planted in Mexico, arguing that the country could no longer meet the national demand for corn because it had resisted using transgenic corn. Monsanto of course said nothing about how the introduction of genetically modified corn would also threaten the sixty-seven varieties of indigenous Mexican corn and potentially destroy the country's corn diversity.[41] Not just the different tastes that come from Mexico's wide array of indigenous corn would be compromised by cross contamination with genetically modified corn, but the diverse characteristics that help make each species of corn more resistant under different conditions. It is these very qualities that make Mexican corn especially important to scientists as they work at developing hybrid corn varieties resistant to climate and environmental change.

There is another side to how free-market forces direct and shape the global food system: growing speculation on food derivatives. A large part of why the price of Mexican corn rose so dramatically and quickly was speculative corn trading. Indeed, speculative trading in commodity futures worsened the global 2007–2008 food-price hikes. With the fallout around financial derivatives since the 2007 global financial meltdown, food commodities speculation is becoming the new commodity in futures trading. One of the banks responsible for the meltdown—Goldman Sachs—is ironically also behind the growing trend in food speculation.[42] In 2010, Goldman Sachs reaped $1 billion dollars in profits from speculation on food derivatives.[43] Food is included in the Goldman Sachs commodity index funds, which basically allows for bets to be placed against the price of staple foods. The role that long-only index fund speculation plays in the commodity market is certainly controversial, and it highlights an ideological clash: speculation versus the regulation of prices.

In testimony to the U.S. House of Representatives Committee on Agriculture in 2008, Scott Irwin, professor in the Department of Agriculture and Consumer Economics at the University of Illinois at Urbana-Champaign and Laurence J. Norton Chair of Agricultural Marketing, explained that after World War II, the price of U.S. grain futures skyrocketed, and President Harry S. Truman declared that "the cost of living in this country must not be a football to be kicked around by grain gamblers." Truman responded to the great increase in prices by demanding that the Commodity Exchange Authority, a precursor to the Commodity Futures Trading Commission, raise futures exchange margins to 33 percent on all speculative positions, going on to say that "if the grain exchanges refuse the government may find it necessary to limit the amount of trading." In 1958, the U.S. Congress banned trade in onion futures in an effort to curb extreme speculative price activity. Similar price increases between 1972 and 1975 prompted the U.S. government to introduce what Irwin described as "drastic measures"—namely, federal price controls along with an embargo on soybean exports.[44]

Irwin brought to the committee's attention the connection between speculative activity and commercial hedging from market risks in the recent past, explaining that the price hikes cannot be neatly blamed on long-only index fund trading. For instance, 500,000 contracts were sold in 2008 to commercial firms involved in corn production and processing, compared to the 250,000 contracts bought by speculative corn traders. In the first quarter of 2008, the "position of short hedgers" was "slightly less than 6 billion bushels." Irwin concluded that "increases in long speculative positions tend to represent speculators trading with hedgers rather than speculators trading with other speculators."[45]

Yet what Irwin does not factor into his assessment is that speculators and hedge funds bet on the market together. In other words, they are part of the same system of free-market capitalism, so separating speculative activity from hedge fund activity is an artificial distinction that distorts the picture of how price fluctuations in the free market work. The jitteriness of the corn market from 2010 to 2011 suffices to make my point. In September 2010, the price of corn was at $4.50 a bushel, increasing steadily to $7.40 a bushel by March 3, 2011. From early to mid-March 2011, the price fell by $1 a bushel when, after the earthquake in Japan, speculators became nervous that the Japanese demand for U.S. corn would drop, prompting speculators, index, and hedge funds to quickly withdraw from the market, which caused prices to drop accordingly.[46]

Where the politics of a rights-based agenda is committed to developing programs and institutional frameworks that will secure and create individual entitlements, the technical and scientific approach to dealing with food scarcity attends to the development of new machinery, more climate resilient seeds, and new agricultural management strategies. Neither, however, fully addresses the economic life of the free market and how these forces configure trends in food volatility. The reason why is that there is an inbuilt bias that favors individualism and privatization in the logic of neoliberalism; this bias presupposes that individual farmers will choose to convert to conservation agriculture or that the dietary choices of meat eaters will shift to a vegan or vegetarian diet or that more corporate governance of the various aspects that make up the global food system will solve the problems of food scarcity. But food scarcity is a *common* problem that needs *common* solutions that try to address the structural distortions arising from speculators who push up the price of food staples, the subsidies that encourage farmers to divert their crop fields to ethanol crop production, the free-trade policies that reconfigure the socioeconomic landscape of the global food system, the weakening of biodiversity, and the corporate entities that gradually increase their monopoly over life's systems.

6

ANIMAL PHARM

I once asked my four-year old daughter while she was drinking a glass of milk: "Where do you think milk comes from?" I wasn't quite sure what to expect because she was notorious for her left-of-field answers to seemingly obvious questions. I must admit even I was surprised when she announced: "A truck, of course!" I was immediately struck by how quickly she had hit the nail on the head. She hadn't been fooled by bedtime stories of farmers wearing checked shirts and straw hats, cows happily nibbling green grass in open fields, chickens wandering around the yard as the sun shines, pigs rolling in the mud, a two-storied farmhouse with its long porch filled with white rocking chairs and a cat asleep on the front steps. She knew exactly where milk came from, and it was most definitely not the fairytale picture of "Once upon a time in rural America . . ."

Ten billion land animals are raised annually in the United States for meat, milk, and eggs,[1] and this gargantuan production cannot be achieved without generating hybrids of machine and organism, or what Donna Haraway so aptly describes in her "Cyborg Manifesto" as the "cyborg machine."[2] Haraway developed her theory of cyborgs to counter dualisms that rely on naturalized oppositions between men and women or nature and machine, proposing that politics demands we move beyond such dualistic thinking and the logic of dominance on which it is presupposed in order to realize fully the liberating potential of the cyborg. In response to Haraway, we need also to be mindful of the power relations distributing the cyborg throughout the social field and the hierarchies of power that such distribution processes create. Can this image of the cyborg retain its political position of disagreement within the status quo as the connection between livestock production, climate change, and environmental degradation enters mainstream politics? In other words, it is one thing to recognize that 18 percent of worldwide GHG emissions can be attributed to livestock production, the proposed solution to which is cultural (eating fewer animal products) and practical (more intensive farming), but this recognition does not address the myriad levels of violence in operation throughout the system of livestock production and the biopolitical configuration of the free-market economy.[3]

A politics might arise from the uncertainty that the image of the cyborg presents to human society, for the industrial food complex represents the antithesis of cyborg disagreement as it disciplines and regulates the creative pulse of material life and living labor, placing it in the service of capital accumulation. It does this in many ways. It plugs the bodies of animals into machines as a way to dominate them. It homogenizes land practices around the system of mass production. It has a monopoly over the reproductive cycle of animals raised for food. It forces farmers into the cycle of credit and debt, turning them into a compliant workforce. It trades workers rights against immigration law. And it institutionalizes food into an oppressive political arrangement that diminishes the creative commons that emerges as people come together to prepare food and share a meal together: the creative combination of color, texture, smell, and taste; food preparation as an expression of care and love for others; the rituals surrounding the decoration of the table; and the thanks collectively given by all at the table in appreciation for the food they are about to eat.

In the industrial food complex, there is no *common life* to be celebrated as the modernization of food production homogenizes how birth, life, nourishment, and death are encountered. Indeed, in this complex the commons is accessed through the political economy of living labor (animals and workers) and the biopolitical economy of material life. We can borrow an observation from Michel Foucault in "Society Must Be Defended" to see how the industrial food complex exercises the sovereign "power to 'make live' and 'let die.'"[4] For this reason, although it is important to recognize, as Foucault did in the first volume of the *History of Sexuality*, that biopolitics marked the onset of modernity, changing how the nation-state exercises power by turning life itself into the object of political control—an idea that Hardt and Negri put to work when they transform the Marxist struggle between labor and capital into a struggle between the common and capital—I would also like to urge some caution over the way in which the concept of biopolitics is put to work.[5]

I share Foucault's position that the site of political struggle is ontological, but I would like to revive the Marxist problematic of how that struggle takes place. As I argued in chapter 3 in my discussion of the "population bomb" thesis as it is used in climate change discourse, it is politically important that we do not water down the struggle between capital and labor because today capitalism has managed access to the commons through reproductive and productive labor. And the same situation, I argue here, holds true in the animal industrial food complex.

Today four companies control 81 percent of the beef market, 59 percent of the pork market, and 50 percent of the poultry market in the United States.[6] Neoliberal adjustments of the agricultural sector have seen the introduction of technologies and management systems that have placed every aspect of life in the service of capital accumulation. And the result is horrific. Two cases in point are revealed by the now notorious Mercy for Animals undercover videos. One video is of Ohio Conklin Dairy Farm operations in 2010.[7] It documents some of the most extreme forms of abuse cows can be submitted to daily: being beaten by crowbars and stabbed by pitchforks, having their tails twisted until they snap, being kicked, being punched in the head, and so on. The other video is of Quality Egg Production in New England, one of the largest egg producers in the United States.[8] It was shot between 2008 and 2009 and documents the cruelty that egg-laying hens encounter on battery farms: living in cages alongside rotting carcasses of other birds, trapped in their cage wire so they cannot access their feed or water, unable to spread their wings, suffering bloody open wounds and broken bones, being thrown live into trash cans, and dying a slow painful death after workers have grabbed them by the neck and twirled them in circles.

I am certainly not the first and I will probably not be the last to raise the red flags on the exploitative and violent treatment of livestock and birds. There is mounting criticism of animal cruelty, with debates over animal rights and liberation falling into one of the following three categories: some argue that the current situation is the result of treating animals as commodities; others connect patriarchal forms of violence to animal cruelty; and some maintain that we just do not seem to recognize that animals also have moral worth. None of these positions is wrong, and they all inform the argument that I am about to put forward. That said, however, the political trajectory they all offer is veganism. Modifying individual eating habits is understood to be either an act of solidarity for the plight of animals raised for food or an act of protest against the institutionalized violence perpetrated against them or both.

The dietary approach also invokes a second-wave assumption that the "personal is political." But how political can the personal be? Does this view run the risk of relying on a neoliberal assumption that promotes a privatized approach to politics? Advocating that individual consumers change the way they eat in protest against the cruelties that animals raised for food are subjected to is certainly understandable. I am a pescatarian who buys local and organic produce as much as possible and who chooses to eat vegan at least three times a week, and I do so because I also want to remove myself from the violence associated with the industrial food complex. I am certainly implicated in the criticisms I am raising here. That said, I continue to be skeptical over how effectively my "personal" eating habits can lead to institutional change. I also have no illusions over the privileged nature of the dietary and food "choices" I can make each week. Not only am I in the advantageous position of being able to afford the increased cost of healthy food, but I am also fortunate enough to live in an area where there is an abundant supply of these foods offered at a variety of retail outlets.

Given the number of food deserts throughout poor African American neighborhoods in U.S. cities, many people cannot afford or even access healthy produce in their local neighborhood. In a food desert—an area suffering from a lack of access to grocery stores with healthy food options—if one does not own an automobile, healthy and affordable food options are rare, and the whole notion of individual choice is held to ransom by a number of fast-food outlets that tend to set up business in poor neighborhoods and whose food has a high-fat, sugar, and salt content. I therefore think we have to be very careful about conflating identity politics with political change.

Nancy Fraser has correctly pointed out that a cultural change in *mentalités* does not necessarily translate into structural and institutional change. Indeed, she shows how privileging the cultural project of identity politics over measures to counter poverty and redistributive justice has had the unfortunate consequence of serving the interests of neoliberal capitalism.[9] Throughout the latter part of the twentieth century, the liberal focus on choice as an expression of individual freedom was put to work in crafting a political position that promoted individualism and personal responsibility in support of trickle-down economic policies. This connection has since justified massive cutbacks in government spending, the end result being that social welfare programs have been curtailed and public assets sold.

Part of the problem, as Fraser sees it in her analysis of second-wave feminism, is identity politics, which has resulted in the subordination of "socioeconomic struggles to struggles for recognition."[10] After pushing the redistributive aspect of the feminist emancipatory project into the background, feminism has been ill equipped to tackle head on the astonishing inequity and poverty that has occurred since Ronald Reagan and Margaret Thatcher first began instigating an unregulated global free-market economy.

Figures on the distribution of wealth in the United States help place Fraser's point in perspective. The term *wealth* refers to the marketable assets a person or family owns minus their debts. Economist Edward Wolff notes that the wealthiest 1 percent of U.S. households owned approximately 38 percent of total wealth in 1998, with the bottom 20 percent of households having zero wealth.[11] Since the Great Recession at the beginning of the twenty-first century, this distribution ratio has not radically shifted. That said, for 2007 the financial wealth of the top 1 percent was approximately 8 percent greater (42.7 percent) than their previous wealth in 2001 (34.6 percent). The term *financial wealth* refers to people's net worth in owner-occupied housing minus their net equity. Even more interesting is that in 2008 for the 13,480 individuals/families earning $10 million a year, only 19 percent of that amount came from wages and salaries.[12] Put differently, the more wealth a person has, the less that person needs to work for his or her income.[13] Meanwhile, negative net worth increased from 15.5 percent in 1983 to 17.6 in 2001, falling slightly to 17 percent in 2004.[14]

If we are serious about realizing a liberating project in the twenty-first century, there is an urgent need to start with a critique of neoliberal capitalism and then use this critique as a way to interrupt how the biopolitical economy and cultural *mentalités* inform one another. In this regard, combining the work of feminist animal-liberation advocates, such as Marti Kheel and Carol Adams, with that of those who focus on the political economy of meat eating, such as Bob Torres, is one way to expand upon and politicize the decontextualized arguments of leading analytic philosophers in the field, such as Peter Singer and Tom Regan.[15]

ANIMAL RIGHTS AND LIBERATION: THE DEBATE

Peter Singer has taught us that human beings discriminate against animals on the basis of their species membership.[16] He argues that the human preference to favor the interests of its own species ahead of the interests of other species is *speciesist*. In this way, the violent treatment of cows, hogs, and poultry is no different than other forms of discrimination, such as sexism or racism. Meanwhile, Tom Regan argues that animals have a moral right to be treated respectfully and that this right is not open to utilitarian negotiation. Regan's egalitarian position leads him to conclude that animals have an inherent value, and so they deserve to be treated in a way that is respectful of this value.[17] Singer and Regan thus offer compelling arguments in support of the moral value of animals. They prompt us to ask whether the interests and rights of animals are en par with those of humans. If their interests and rights can be proven to be equal to that of humans, as both Singer and Regan argue in their respective ways, then a valid argument in favor of human obligations toward nonhuman animals can be made. The moral injunction of vegetarianism stems from their shared conclusion that animals deserve moral considerability. For Singer, because animals feel pain, they are morally considerable. Meanwhile, Regan argues the same thing on the basis that animals have the capacity for self-awareness.

The limitation of Singer and Regan's position, however, arises from the way their arguments presuppose an essentialist subject, one who is somehow free of its contextual determinants. The problem of prejudice or discrimination based on species membership that both Singer and Regan highlight is also at its core ideological. It points to a set of misplaced beliefs and values that fail to understand the inherent value of animals and their interests. Both authors remind us that human beings occupy a position of privilege vis-à-vis animals, and they offer tight arguments to counter the speciesist character of human nature. Yet their analytic methodology fails to recognize fully the social relations that structure how privilege is put to work.

A quick thought experiment suffices to make my point. Sue and Tom are dog "owners." Sue treats her dog as a companion, an equal member of her household. She openly displays affection toward her dog, cuddling up to her at night and playing with her in the doggy park along with other dogs and dog owners. Tom raised his pit bull to fight, intentionally torturing, wounding, and maiming the animal in an effort to develop and fine-tune his fighting instinct. The dog is aggressive, unsocial, and unpredictable. The point of this thought experiment is to highlight the fact that both Sue and Tom are dog "owners," yet the position of privilege they occupy vis-à-vis their dogs has produced very different outcomes. In Tom's case, there is an invisible line connecting masculinity, power, emotional arousal, the violent treatment of animals, and sexist values that shape and inform the culture of dog fighting. The point here is that we cannot neatly extract a subject from its social, cultural, economic, and political circumstances.

In *The Sexual Politics of Meat*, Carol Adams examines the speciesism and misogyny characterizing the content of menus, advertisements, and popular slogans.[18] She argues that female and animal bodies are objectified by a patriarchal value system that is inherently violent. She traces the cultural meaning of meat eating, masculinity, and virility as intertwined phenomena. Just as women's bodies have been commodified and objectified, so, too, have the bodies of factory-farmed animals. The individual life of a woman or animal is rendered abstract as each is objectified. Adams invokes the concept of patriarchy as a gendered system of violence to describe the systems of oppression that both animals and women experience. From here, she asks: "Which images of the universe, of power, of animals, of ourselves, will we represent in our food?"[19] She goes on to conclude that what we eat is a direct reflection of our politics, and for this reason the only truly ethical and feminist position to take is that of veganism.

Meanwhile, Marti Kheel draws attention to the way in which both Singer and Regan devalue personal relationships and affective ties, discounting empathy and care as morally significant criteria in ethical thinking. Kheel counters the masculinist paradigm of competition and independence by closely studying the patriarchal unconscious operating throughout the work of four leading holist philosophers—Theodore Roosevelt, Aldo Leopold, Holmes Rolston III, and Warwick Fox.

Kheel proposes that holism subordinates "empathy and care for individual beings to a larger cognitive perspective or 'whole.'"[20] In contrast to the usual abstractions and rationalisms that typify environmental ethics and positions supporting animal liberation, she advocates an ethics of care and empathy. The important point she makes is that empathy for the individual is overridden when primary value is given to the abstract "whole."

Kheel defines the holist focus on abstract constructs as masculinist. It pits reason against emotion, reorganizing the complex interaction and inter-dependency of life through a hierarchical system of abstract values. She, too, is ethically committed to a vegan diet. For her, this approach provides an effective way to counter grand master narratives that propound mas-culinist views (competition, heroism, rationalization, and abstraction). In another respect, what she is proposing is a corporeal strategy: through the redistribution of the social field's corporeality, the violence inherent to that field is interrupted and reconfigured. I do not disagree with her on this point; I think the strategy has the potential to cut to the core of how politics and ontology have been combined in the biopolitical economy. I would, however, like to see how this approach might be better positioned to pro-duce disagreement in the larger social field and with respect to the objec-tive forms of violence endemic to the political economy of capitalism.

Bob Torres has published a terrific analysis of the political economy of animal rights, *Making a Killing*, and I certainly sympathize with the social anarchist position he presents. He states from the outset of the book that sexism, racism, and our relations with animals are structured through a series of historical relations of domination that benefit one group of people or one species over another. Using Marxist theory to highlight the ideo-logical and economic relations behind livestock production, Torres situ-ates animal rights alongside gender, race, and class oppression. He draws striking parallels between the dehumanization of nonwhites, common to discussions over racism, and the "treatment of animals as mere things."[21]

The main thrust of Torres's argument is that capitalism has commodi-fied animals. He maintains that animals produced for food are "superex-ploited living commodities" because they are the property of someone, they are commodities, and they produce commodities to feed the produc-tive labor power of the capitalist system.[22] In his conclusion, he, like Kheel and Adams, explains that the most effective way to contest animal exploita-tion and commodification is by becoming vegan. The corporeal nature of his politics, like those of Kheel and Adams, comes from using the affective potential of food in its connection with bodies and material life to chal-lenge the dominant relations of power that manage the social field and common life.

So I return to the question of whether modifying one's personal eating habits is enough to produce change at an institutional and structural level? Does this approach sufficiently engage with the objective forms of violence endemic to the industrial food complex and the inequities propounded by that system? More troubling, does equating politics with personal responsibility in this way unwittingly reinforce a basic constituent of neoliberalism—that individuals, not governments or historical forces, are personally responsible for their own successes and failures? One might easily see why Singer and Regan's analytic approach is susceptible to this blind spot; the implication of their essentialist subject is that it is freed of its contextual limitations—a subject who is accordingly an independent entity in the world. But the situation with respect to the ecofeminist and anarchist vegans is not so clear-cut.

What is key here is that ethical food choices cannot be separated from the material conditions determining food production and modes of subjectification (race, class, gender, species). For instance, most vegans choose to eat soybean products.[23] Yet the transnational politics of soybean production is responsible for the razing of large parts of the Amazon rain forest that is facilitating the institutionalization of North–South power relations. Further, the principle at the core of veganism—the individual's power to choose and take responsibility for what he or she consumes—has unfortunately already been co-opted by neoliberal capitalism in its principles of individualism and competition. As a politics of consumption, the vegan approach runs the risk of facilitating the culture of consumption that capitalism advances. In addition, the politics of veganism needs to move beyond the personal identity politics of individual food choices and the factionalism this view promotes by encouraging vegans to form alliances with other activist groups regardless of whether the individuals who make up these other social and environmental justice movements are vegan or not. The case of Lierre Keith immediately springs to mind.

Keith was a vegan for twenty years but found the diet unhealthy, prompting her book *The Vegetarian Myth*, in which she provides a critical and fascinating examination of the history of agriculture.[24] At the fifteenth annual Bay Area Anarchist Book Fair in San Francisco, a group of vegans accused Keith of being an animal holocaust denier, and thirty minutes into her book presentation, ironically right after she had announced we should not eat factory-farmed meat, they attacked her with chili pepper–laced pies that left her with sore eyes for a few days after the incident.

We need to push the structuralist analysis beyond a critical analysis of culture and personal identity politics. This task is both practical and theoretical. Practically speaking, it requires solidarities to form across disparate groups so that they are better positioned to engage critically with the material conditions of life and the biopolitical economy of material life. Meanwhile, the task for political theory is to use the struggle between the commons and capital that Hardt and Negri describe to inflect analyses of the struggle occurring between labor and capital without necessarily supplanting the latter with the former.

By way of a starting point, I propose one qualification to Torres's argument: the biopolitical economy of animals raised for food and of the laborers who work within this industry arises less out of the commodification of their labor and more from the contradictions central to the process of capital accumulation. Torres overemphasizes the commodity form in his analysis of political economy, and so he neglects an important distinction: capital is not a thing; it is a process. As a process, capital needs to stay in circulation; when it encounters potential blockages, it goes into crisis. Capital then appropriates the crisis as a way of transcending it. This appropriation then opens up new avenues for capital accumulation, which is where Foucault's concept of biopolitics is incredibly helpful. This concept encourages us to examine how capital regulates common life and the mutual dependencies that constitute it, rendering it noncommon in the process. In addition, it is labor that determines how the commons is accessed. For this reason, the Marxist antagonism between labor and capital remains.

Marx recognized that when capital hits upon a limit, it either circumvents or appropriates it in some way. In the *Grundrisse*, he remarks: "But from the fact that capital posits every such limit as a barrier and hence gets *ideally* beyond it, it does not by any means follow that it has *really* overcome it, and, since every such barrier contradicts its character, its production moves in contradictions which are constantly overcome but just as constantly posited."[25]

So what might be some of the potential blockages that the industrial food complex presents to capital? There are six: (1) production (leading to the uniformity of production); (2) labor (producing a compliant workforce through systems of surveillance and the securitization of material life); (3) risk (reorganizing and managing material life); (4) reproduction (regulating the material limits of a body and life); (5) knowledge (the privatization of public research); and (6) demand (reconfiguring taste and appetite). All of these strategies rely on separating the means from the end, objectifying life, and thereby managing every aspect of life, from sex to pregnancy, motherhood, lactation, sustenance, and death. This is the story of material life being reorganized into predictable, manageable, self-contained units of space and time.

PRODUCTION

After World War II, the strength of the meatpacking unions resulted in a master contract that meant by 1960 the average meatpacker's wage in the United States was 15 percent higher than the average manufacturing wage.[26] A serious limit to capital thus lay with the trade unions that oversaw the institution of basic labor rights throughout the meatpacking industry, which in turn increased the cost of labor. However, by 2005 the median annual salary of meat and poultry workers had plunged to $21,320, more than one-third less than the median $33,500 that workers in all other manufacturing industries earned.[27] The industry overcame the rising cost of labor by deskilling the labor force, improving labor productivity, and vertically integrating the disparate elements of animal food production. The result was the fast-paced, highly organized, cruel, and dangerous system of the animal–industrial complex.[28]

As the farm was industrialized, animals were moved from the outdoors to warehouse facilities, where there is no natural lighting, living conditions are cramped, disease and depression are high, and life is short. To stop birds pecking each other, farmers remove anywhere between one-third and one-half of their beaks; the pain causes some birds to stop eating, and they quite simply die of starvation. Broiler chickens live in chronic pain for 20 percent of their life, and their bones are so weakened as a result of their confinement that when they are removed from their cages, their bones are quick to snap. Male chicks born to egg-laying hens have no economic value and are either gassed, macerated alive, or thrown into trash cans. Piglets, weaned just ten days after being born, as compared to the natural thirteen weeks, are left with a longing to suck. They satisfy this urge by chewing on the tails of other piglets. The average natural life of a cow might be twenty years, but when the milk production of a dairy cow decreases at around five to six years, she is sent to the slaughterhouse. And the list goes on.

In addition to the cruelty animals are subjected to, workers in the animal–industrial complex are also at risk. Health and safety issues for workers abound. In 1990, the Health Hazard Evaluations of two poultry plants processing more than four hundred thousand birds daily, conducted by the National Institute for Occupational Safety and Health, found that 20 to 36 percent of workers had work-related cumulative trauma disorders. An Occupational Safety and Health Administration study from 1989 found that poultry workers were required to make more than ten thousand repetitions per shift, causing serious repetitive-motion problems.[29] The U.S. Government Accountability Office noted in 2005 that although injuries and illnesses in the meat and poultry industry had fallen from 29.5 injuries per 100 full-time workers in 1992 to 14.7 in 2001, the recorded rate was still one of the highest of any industry.[30] Others have pointed out that these figures present a distorted picture because many injuries and illnesses go unreported.[31]

Workers on the slaughterhouse floor are kicked or bitten by frightened and maimed animals; they are physically injured by the knives they use and from the repetitive tasks they perform; and they experience psychological trauma as a result of participating in and seeing mass slaughter on a daily basis. All of these factors provide incentive for workers to perpetrate extreme acts of cruelty against the animals.[32] One male worker at Morrell hog slaughterhouse explained,

> Another time, there was a live hog in the pit. It hadn't done anything wrong, wasn't even running around the pit. It was just alive. I took a three-foot chunk of pipe—a two-inch diameter pipe—and I literally beat that hog to death. Couldn't have been a two-inch piece of solid bone left in its head. . . . It was like I started hitting the hog and I couldn't stop. And when I finally did stop, I'd expended all this energy and frustration, and I'm thinking, what in God's sweet name did I do? . . . People go into Morrell expecting respect and good working conditions. They come out with carpal tunnel, tendonitis, alcoholism, you name it, because they're under incredible pressure and they're expected to perform under intolerable conditions. Or they develop a sadistic sense of reality.[33]

The prevalence of intentional torture and abuse of the animals by slaughterhouse workers has prompted researchers to analyze the relationship between the violent nature of slaughterhouse work and high incidences of crime, alcoholism, drug use, and domestic violence that characterize slaughterhouse communities.[34] For instance, Amy Fitzgerald has found the variables of unemployment, social disorganization, and demography that typify slaughterhouse communities do not fully explain the rise in "total arrests, arrests for violent offenses, arrests for rape, arrests for sex offenses, the arrest rate scale and report rate scale" in these communities.[35] In her sociological assessment, the violence endemic to slaughterhouse communities is an expression of the psychological distress slaughterhouse workers experience at work.

LABOR

Given the low pay, physical risks, and psychological stress workers encounter in the meat, poultry, and dairy industry, it is unsurprising that the industry has high rates of worker turnover. This turnover creates another crisis for capital: a labor shortfall. Along with deskilling the workforce and making meat and dairy production more efficient, the concentration of the meatpacking industry in rural areas has isolated that workforce from strong urban-based unions, allowing companies to seek out a more vulnerable workforce. It has targeted those who have few employment opportunities and are reluctant to collectivize, more dependent, and easily intimidated: immigrant labor.

Meatpacking facilities were historically located at railroad terminals in urban areas close to where cattle arrived. In 1961, this situation changed when Iowa Beef Processors, Inc. (now Tyson Fresh Meats) located its meatpacking facility in close proximity to the feedlots in rural northwestern Iowa. Transportation costs were lowered as cattle no longer had to be transported to urban centers to be slaughtered and packed. Further, Iowa Beef began boxing meat close to the feedlot, which further lowered transportation costs because fat and bone were removed in preparation for packing. More significantly, union activity was weak, if not nonexistent in rural areas, which meant the master contract would no longer be enforced. The problem then became one of how to keep the unions out, and companies realized that the best way to achieve this goal was to employ migrant workers.

The North Carolina Company Police Act of 1991 (N.C. Gen. Stat. §74E-1) allowed companies to employ company police and guards who are "empowered to carry weapons, make arrests, and pursue 'suspects' off company property as long as an incident began on company property."[36] In 2002, the U.S. Supreme Court decision in *Hoffman Plastic Compounds, Inc. v. NLRB* (535 U.S. 137 [2002]) ruled five to four that immigration law takes precedence over labor law. More specifically, the ruling held that because of the illegal status of undocumented workers, these workers were not able to turn to the U.S. courts and labor laws to seek back pay for lost wages after being illegally fired for union organizing. The combination of the Company Police Act and the *Hoffman* decision was lethal for U.S. meat and poultry workers. Undocumented workers were stripped of their rights, and the rights of other minority workers were further compromised. With immigration law trumping labor and human rights law, the floodgates for the intimidation of immigrant employees were opened. Recruiters went to great lengths, some of them illegal, to employ immigrant workers, and employment of undocumented workers in the meat and dairy industries increased.

In December 2000, the U.S. Immigration and Naturalization Service raided a Nebraska Beef plant. More than two hundred workers were deported. In 2002, because all the witnesses had been deported, a federal judge dismissed the indictment of company managers and administrative staff for a criminal conspiracy to recruit and transport workers from Mexico after providing them with false documents to work at Nebraska Beef.[37]

Using the Company Police Act of 1991, security forces at Smithfield Foods received "special police agency" status in 2000. On November 14–15, 2003, when immigrant workers protested the dismissal of coworkers by walking off the plant, they were assaulted and arrested by plant police. The case went to trial in September 2004. The packing company police and guards were found to have "physically assaulted employees exercising their rights" and to have "threatened employees with arrest by federal immigration authorities."[38]

As studies have shown, an added benefit for companies who use immigrant labor is that these workers tend to underreport injuries and work-related illnesses because of "language barriers, workers' fear of losing their jobs, workers' concerns about immigration status, incentive programs that reward low rates of absenteeism, and lack of access to health care."[39] However, for even those workers who want to seek out medical advice, there is the added problem of geographical isolation. Health-care resources are already stretched thin in rural areas. Rural social services are placed under further strain when there is an influx of new residents; many speak English as a second language.

By isolating the labor force in rural areas and employing minority workers, the industry broke the strong arm of the meatpackers' union. Employing workers from minority groups made for a more compliant workforce that was less likely to collectivize, complain, or quit their job. In this way, capital had overcome another serious barrier: the right of workers to collectivize and petition their employees for improved wages and working conditions. From here on, the market, not unions or government policy, set the wages of workers in the U.S. meat and dairy industries.

RISK

Farming has historically been a volatile business. It depends on ecological cycles that are notoriously unreliable and difficult to control—a typical case of what Hardt and Negri might describe as the commons coming into conflict with capital. A classic example is the U.S. Dust Bowl of the 1930s, when millions of acres of farmland were lost after years of prolonged drought and relentless dust storms. Unpredictability of this kind puts capital into crisis. As I outline later, this crisis is overcome by accentuating the struggle between labor and capital.

During the 1970s, domestic and international sales of U.S. farm products were booming. From 1970 to 1973, net farm income grew from $14.4 billion to $34.4 billion.[40] Trade barriers had been lowered, and the USSR was buying up U.S. grain. The sector needed to increase productivity and was encouraged to modernize and become more efficient. To meet growing demand and remain competitive, farmers borrowed heavily, speculating that their revenues would remain strong. But alarm bells were sounding on rural America's horizon. In addition to the debt farmers took on, they paid inflated prices for land. From 1957 to 1977, the average per acre price of farmland grew 364 percent.[41] After the restrictions on Federal Land Bank lending were lifted, a long period of upward growth in U.S. land prices ensued, attracting the attention of investors until a speculative bubble emerged (not dissimilar to the real-estate bubble that led to the economic meltdown in 2007).

The situation changed as demand for and the cost of U.S. farm products declined. On January 4, 1980, President Jimmy Carter embargoed U.S. grain products to the Soviet Union as a putative measure following the USSR's military occupation of Afghanistan in December 1979. The global recession from 1981 to 1982 hurt the U.S. farm-export industry as agricultural exports fell 20 percent from 1981 to 1983. As a result, farm income fell. When interest rates rose, farmers were left with high loan repayments, and farm foreclosures grew. Faced with financial ruin, U.S. farm communities were hit hard, as attested to by the rising rates of divorce, alcoholism, and child abuse.[42] The end result was what is now referred to as the "U.S. farm crisis" of the 1980s.

Farm foreclosures hit small farmers hardest, providing fertile ground for larger investors to capitalize off their hardship. Farmers were either driven from their land or forced to consolidate. This vertical integration of the food system changed farmers from independent operators and an important ingredient in the local food system and economy into passive subjects more dependent on the global market and contract farming.

As company contractors, farmers are obligated to meet the quality, quantity, and production timeline targets set by the company. In addition, they are often obligated to use specific methods set out by the contracting agent, and they forego the power to decide how animals are raised. For instance, farmers growing birds for Purdue Inc. have been forced to make the bird production process more efficient and to follow the production guidelines set by Purdue, which has resulted in birds being permanently located inside large henhouses without access to sunlight or fresh air. In addition, ownership of the farm no longer guarantees farmer independence as the company-owned contract-farming system shifts financial and production risks away from the company and into the hands of the farmer.

Consider the average poultry farmer who holds a contract with Goldkist or Purdue or Tyson. The company owns the birds and feed and provides the farmer with transport and medicine. Contracts are issued on a flock-to-flock basis. The farmer is responsible for the high cost of birdhouses (each in excess of $300,000), irrigation, feed, lighting and ventilation systems, as well as waste disposal. He or she (many are women) incurs the costs of all upgrades demanded by the company without any long-term guarantee from the company that it will continue to purchase his or her birds. Heavily indebted, the farmer is completely dependent on the company, and reports of company retaliation against farmers who complain or protest against the terms of their contract abound throughout the popular media.

Because dairy, hog, and poultry farming is capital intensive, most of this sector is leveraged. And one of the main reasons why farmers seek debt financing is to purchase real estate, which includes building, livestock, poultry, and grove development. In 2007, three-fifths of farmer debt were for real estate, with financing operational costs, machinery, and equipment following close behind. After a $26 billion increase, for 2007 total U.S. farm debt reached $240 billion. Interestingly, the number of farms using debt financing that year had fallen to 31 percent from 60 percent in 1986, but the debt in question was more "concentrated in fewer, larger farm businesses," which seems to support my thesis that along with vertical integration comes individual farmer debt accumulation.[43]

REPRODUCTION

Another hurdle for capital came from the material limits of the reproductive body. As such, today the sex life of animals raised for food has been reduced to data, outputs, and statistics. Lactation, weaning, sex, pregnancy, birth, and the vital combination of energies, affects, and fluids have been transformed into an informational body whose material processes are mediated by capital. Put differently, the effects (offspring, food, and so on) and processes (care, mothering, nurture, and feeding) of intimacy have been placed in the service of capital accumulation. In this regard, the feminization of labor that Hardt and Negri identify in their analysis of the biopolitics of capital and the technical composition of labor specific to this phenomenon stretches across species.[44]

Breeding-management programs keep detailed and timely records of animals' reproductive cycles. In the move to make reproduction more efficient, there has been a shift away from "natural" to "artificial" reproductive technologies. In other words, surplus value arises from the biopolitical economy of immaterial labor as affects, semen, and knowledge work in tandem with the commodification of intimacy.[45] Artificial insemination has become a popular breeding-management technique. Sows can be artificially inseminated with semen collected from boars trained to mount an artificial sow. This system produces more uniform pigs and lowers production costs because fewer boars are used. In comparison to the old system, where all sows in heat were moved to an area where they were impregnated by the boars, artificial insemination takes less time. If the system is managed carefully, conception rates improve and resulting farrowing rates (birth of piglets) are strengthened when sows are artificially inseminated.

The management of the reproductive cycle does not end here. During the final stages of gestation and sometimes for her whole pregnancy (approximately 115 days), the sow is contained in a farrowing crate, and then once she gives birth, she is confined there until her piglets are weaned. Slightly longer and wider than the sow, a farrowing crate severely restricts her movements (she basically cannot turn around), which leads to muscular and joint weaknesses and sooner or later impaired mobility. Denied her innate behaviors, such as rooting and nesting, she suffers from chronic depression and frustration and develops behavioral disorders such as stereotypies (obsessive compulsive behavior) as a way to cope. The piglets access her breasts from a slatted floor, through which their feces and urine can pass outside the mother's crate. Apart from higher rates of piglet survival (the restricted movement stops the mother from lying on the piglets), another reason why farrowing crates are used is to maximize the use of space.

In 2000, U.S. milk production had increased 45 percent since 1975.[46] Statistics show that just for the month of August in 2010, milk production in twenty-three major U.S. states totaled 15 billion pounds, and the milk produced per cow averaged 1,796 pounds.[47] Large-scale production of this kind is possible only because of widespread neoliberal structural adjustments to the animal agricultural industry. In place of the old farm model, where there were a variety of animal species and crops grown, today farmers have been forced to specialize their production. Measures have been introduced in order to improve efficiency (less labor), production, and output. They involve replacing laboring bodies (humans and animals) with new machinery and equipment such as milking machines, waste-handling equipment, and advances in milk storage and refrigeration, as well as with pharmaceuticals such as recombinant bovine somatotropin hormone treatment,[48] which is used to increase milk production, and new designs for animal "housing" and new feed systems.

Animals that are too old to reproduce and lactate are typically sent to the slaughterhouse, but with advances in scientific research the natural limits of the reproductive cycle are made more productive. In 1996, scientists produced the first cloned mammal—Dolly the sheep—a method that allows the limits set by the life cycle to be extended.[49] In 2001, a group of scientists under the leadership of Steve Stice from the University of Georgia at Athens used the genetic makeup of a cow that had grown too old to reproduce to clone eight full-term calves. The Steve Stice Lab proudly announces under the heading "What's Hot in the Stice Lab" on its Web site that it was the "first to produce a clone from an animal that had been dead for 48 hours," highlighting that this procedure opens new opportunities not only for preserving endangered species, but also for developing agriculture as well.[50] Proudly showcasing its power to make life and let die, the lab proclaims that it has produced fifty cloned calves and one hundred cloned pigs this way.

That said, cloning has a low success rate because many cloned animals suffer from immune deficiencies or organ failure. Further, as noted by S. M. Willadsen and associates, cloning calves by nuclear transfer can lead to large calves. "Large-offspring syndrome," as this phenomenon is otherwise called, not only endangers the health of the cloned animal but also puts the surrogate mother at risk.[51] Yet in January 2008 U.S. Food and Drug Administration (FDA) scientists issued a report that strangely prompted the FDA to conclude "that meat and milk from cow, pig, and goat clones and the offspring of any animal clones are as safe as food we eat every day."[52] Despite the FDA's confidence, peer-reviewed research paints a more uncertain picture.[53]

KNOWLEDGE

The new technologies, pharmaceuticals, and management systems used in animal agriculture are the direct outcome of neoliberal research and development policies extending back to the early 1980s. The 1980 Bayh–Dole Act (Pub. L. 96-517, Patent and Trademarks Act Amendments), introduced during the Reagan administration, facilitated strong alliances between academia and commercial sectors.[54] In this way, the democratic potential of science—science by and for the people—was privatized. It is no surprise that Stice juggles two roles, one as a public researcher in a public institution and the other as an entrepreneur. Fifty-one percent of his time is as a scientist at the University of Georgia, a public higher-education institution, and 49 percent with ProLinia, a biotechnology company established in 2003.[55] Bayh–Dole expedited the commercialization of federally funded university research.[56]

As the public sector–private sector alliance prospered, so too did a high-risk investment culture, both of which were in turn supported by neoliberal forces that had created a deregulated and finance driven economy that fed the alliance with investment capital. Massachusetts biotechnology firm Advanced Cell Technology Inc. speculated in 2001 that the market for cloned dairy cows and beef cattle might reach $1 billion. Smithfield Foods, the world's largest hog producer and processor, entered into a Technology Development Agreement with ProLinia Inc. in 2000 after providing the biotechnology company with an equity investment of $1 million in support of research and development into cloning pigs.[57] The agreement sought to "commercialize ProLinia's cloning technology by arranging for ProLinia to provide cloned embryos from Smithfield's elite genetics for implantation into sows for gestation" as well as to develop "an embryo transfer Standard Operating Procedure" that would meet "industry biosecurity standards."[58]

On June 30, 2003, ViaGen Inc. announced it had purchased ProLinia along with its contract with Smithfield Foods and its scientific talent, intellectual property, and nonexclusive license to use nuclear-transfer technology (used in animal cloning) for agricultural purposes. ViaGen provides genomics and assisted breeding services and products to the animal agriculture industry. Excitedly commenting on the acquisition, ViaGen's cofounder and president Scott Davis stated: "In the biotech business, it's rare to find a single company that allows you to significantly boost your cash flow, fortify your scientific team and gain access to patent rights in one fell swoop. Acquiring ProLinia brings us all of these things."[59]

DEMAND

A decrease in or stabilization of demand can cause chronic problems for capital. If capital is value in motion, as Marx described it in the *Grundrisse*, then demand is a key ingredient to keeping it moving. Without demand, capital quickly comes to a screaming halt. Two factors influencing demand of animal products are the basic calorie needs of any individual body and changes in consumers' purchasing habits (opting for other products).

In 2002, Americans spent approximately $115 billion on fast food. At the same time, more than 60 percent of adults and 13 percent of children in America were classified as overweight or obese.[60] This situation cannot be solved through more exercise, greater self-control, or working mothers' rediscovering the kitchen again. There are other factors, such as the food deserts throughout poor urban neighborhoods, where only fast food is readily available; the cost of fresh produce as compared to cheap high-calorie food products sold by fast-food outlets, which offer larger portions at lower prices; as well as the aggressive marketing of high-fat food to America's youth through gimmicks such as toys at McDonalds or as part of the U.S. School Lunch Program for poor students.[61] Studies show that food with high sugar or fat content or both is highly addictive.

Not only is fast food unhealthy, but the marriage between technology, knowledge, and food production sometimes makes for a lethal combination. In 1993, Jack in the Box hamburgers were contaminated with an *Escherichia coli* strain that resulted in the death of four children and food-borne illness in 750 people. In 1999, after failing the U.S. Department of Agriculture Hazard Analysis and Critical Control Point Systems test three times in just eight months, Texas-based meat processor and grinder Supreme Beef Inc. was forced to shut down its operations. One test showed nearly 50 percent of its beef was contaminated with salmonella. The company responded by filing suit against the Department of Agriculture in federal court. What resulted was a series of court cases that ended up in the U.S. Supreme Court, which ruled in favor of Supreme Beef when the company argued that salmonella is a *natural* substance and therefore not subject to government regulation and that it is harmless if meat is properly cooked.[62]

In 2002, ConAgra recalled 19 million pounds of beef from its processing facility in Colorado after *E. coli* was discovered in the meat. Today the *E. coli* problem is being solved by injecting meat with ammonia. More recently, in 2010, Purdue recalled more than 90,000 pounds of frozen chicken nuggets after traces of blue plastic were discovered in the meat.[63] Also in 2010, more than 550 million eggs were recalled after 2,000 people became ill from salmonella.[64]

On the one hand, the body has a limit to the number of calories it needs on a daily basis. However, food products with high sugar and fat content overcome the material limit of daily calorie intake by making food addictive and by dulling the innate inclination toward consuming healthy food. Demand also goes into crisis when consumers no longer trust the quality of the food they buy. Michael Pollan succinctly notes: "A diet based on quantity rather than quality has ushered a new creature onto the world stage: the human being who manages to be both overfed and undernourished."[65] Or as Raj Patel points out in *Stuffed and Starved*, the global system of food production and distribution has produced the alarming paradox of 800 million starving and 1 billion obese.[66] It is this contradiction that prompted Patel to conduct a compelling critique of the mechanisms of the free market, later going on to argue in *The Value of Nothing* that there is a real disconnect occurring between the price of food and value. He succinctly summarizes the problem with the following conundrum: if we factor in the human and ecological costs of producing a hamburger, then the real cost of a hamburger would be more in the vicinity of $200.[67]

My daughter's image of milk coming from a seven-axle, stainless-steel milk tank truck roaring along the freeway may at first seem to have nothing in common with the Real California Milk image of a black and white Holstein dairy cow standing in a field of luscious grass with snow-capped mountains in the background. One is the artificial, hard body of reason, science, and technology. The other is the warm, soft body with all its appetites, passions, and vulnerabilities. The political concerns surrounding the animal industrial food complex emerge at the interstices of these supposedly unrelated images, for when taken together they bring to light the harsh reality at the core of capital: capital deals with a crisis by appropriating and placing the contradictions that the crisis poses (sufficient and predictable food supplies, consumer demand for certain foods, investment costs, land, and cost of labor) in the service of capital accumulation.

As frightened animals enter the slaughter process, they understandably kick, bite, shit, and collapse, and their muscles go into spasms. Meanwhile, human male bodies are put to work beating, shooting, stabbing, stunning, and slashing large, heavy, struggling animal bodies. Human female bodies work at quickly deboning, gutting, slicing, and boxing smaller dead animals; they are sexually harassed by male supervisors; and some miscarry on the production line because bathroom breaks are strictly limited.[68] In this context, men, women, and other-than-human animals equally urinate, excrete, and vomit. Bodies develop rashes; they ache, bleed, swell and are soiled, deafened, bruised, broken, and maimed. They are transformed and modified by high-fat and sugar diets and injected with bovine hormones. And the list continues.

These graphic descriptions highlight the political organization of material life: biopolitics. As capital organizes bodily affects, it is also a gendering, racializing (using immigrant labor), impoverishing (deskilling the labor force), and speciesizing process. Sociocultural norms of masculinity—strong, powerful, impenetrable, and muscular—distribute male bodies across the slaughterhouse floor. Sociocultural norms of femininity—detail oriented, patient, and careful—similarly organize female bodies along poultry assembly lines making precise rapid cuts.[69] Legislation inscribes bodies along racial lines and according to citizenship status, facilitating the exploitation of undocumented immigrant bodies. Bodies are put to work consuming the fatty, sugary, antibiotic-laden products of animal pharm, whose addictive properties transform the body's psychic and physical makeup.[70] The breasts of female animals are hooked up to milking machines, their bodies are pumped with hormones so that they lactate excessively, leading to crippling pain as a result of chronic mastitis. The lives of other female animals are reduced to their reproductive functions, living an endless cycle of pregnancy, birthing, and feeding without ever experiencing intimacy with their fellow animals.

Capital organizes bodies into consistent subjects along the lines of class, race, gender, citizenship, or species. Capital spatially distributes these bodily organizations by inserting them into different locations: the slaughterhouse floor, the meatpacking conveyor belt, the feedlot, the farrowing crate, the isolated or impoverished setting of rural communities. When these elements are taken together, the entire system consumes 30 percent of the planet's land surface.[71] Through a process that distributes, manages, and organizes the sensible field of material life, the unpredictable nature of bodies is rendered habitual.

As Elizabeth Grosz has consistently argued in her work on feminist corporeality,[72] bodily matter is a lively combination of organs, flesh, nerves, liquids, and bones, and as these elements form connections over time, they affect and are affected by their milieu. From here, habits form, memories take hold, and libidinal energies set into play a variety of material combinations. Throughout the animal–industrial complex, this organization and distribution of corporeal processes that make up the spatiotemporal configuration of the commons is mediated by capital to further capital accumulation. This is the biopolitical economy of capital. As Hardt and Negri assert, the "distinctive feature" of this kind of affective labor is that "paradoxically the *object* of production is really a *subject*, defined, for example, by a social relationship or a form of life."[73]

People's attempt to solve the problem that livestock production poses for climate and environmental change—including land-use changes from deforestation to pasture degradation; excessive water usage; the production of 37 percent of anthropocentric methane gas, which has twenty-three times the global-warming potential of CO_2, not to mention ammonia emissions and nitrous oxide pollution—by advocating that livestock and agricultural production be further intensified but at the same time arguing for cultural changes in individual dietary choices seems to be putting the identity politics of vegetarianism and veganism to work for the wrong reasons. In fact, I fear that the political integrity of vegetarianism and veganism is being completely co-opted by the neoliberal principle of individual choice to further the biopolitical economy and the privatization processes of the free market, all the while masking the violence endemic to the livestock industry.

7

MODERN FEELING AND
THE GREEN CITY

I have visited many cities over the years. Some that left a strong impression include Berlin, Budapest, Chicago, Hyderabad, Kuala Lumpur, Mumbai, New York, Paris, Seattle, and Sofia. I currently live in Cincinnati, Ohio. Since moving to the United States nine years ago, I would have to say that my favorite North American city is Chicago. It is one of the few cities of the world where you can sit on the beach with the skyline immediately behind you or where you can promenade along the waterfront for miles and miles. It is also one of a handful of U.S. cities where bicycling is the norm and public space abounds. The restaurants are hip and alive; its music scene is thriving; its architecture is bold; the art on show is often provocative and cutting edge; and the intellectual scene is stimulating and challenging. In a nutshell, the city has shed its blue-collar rustbelt image, and in its place it has joined the ranks of other global cities the world over.

Chicago is also one of the few cities in the United States that has ranked highly on the various green cities measuring indexes. Over the past decade and under the leadership of Mayor Richard M. Daley (1989–2011),[1] it has undergone a green facelift, moving it into fourth position on the 2008 SustainLane U.S. Sustainable City Rankings.[2] By making *green urban Dev.*

Chicago an "environmentally friendly" city, Mayor Daley expanded on the new and shiny global brand identity that Chicago had forged for itself in a way that reinforced its competitive economic and cultural status within the global arena and in the process generated a greener way of life for Chicagoans. Put differently, as a green city, Chicago epitomizes modern feeling as much as it produces that feeling, and it is this connection that I am interested in exploring further in this chapter. There are obviously many other examples of green urban development that I could have chosen to look at—Abu Dhabi, Tianjin, and Freiburg, to name just a few. Yet in an effort to stave off flattening the theoretical terrain by avoiding the discomfort that can arise from being too close to the issue at hand, I thought it better to focus my analysis on a city that I have a strong attachment to as opposed to one that is by and large present for me as only an abstraction.

I am curious as to how the greening of the built environment is tied to the broader phenomenon of modern feeling on which Fredric Jameson casts a spotlight: "This modern feeling now seems to consist in the conviction that we ourselves are somehow new, that a new age is beginning, that everything is possible and nothing can ever be the same again; nor do we want anything to be the same again, we *want* to 'make it new,' get rid of all those old objects, values, mentalities, and ways of doing things, and to be somehow transfigured."[3]

That said, the appetite for "newness" Jameson describes has the unfortunate consequence of generating amnesia throughout the social field, an amnesia that inhibits institutions and social organizations from effectively intervening in the structural inequities that infuse our common landscape.

Indeed, as David Harvey points out, modern feeling facilitates an axiomatic of capital as it configures and distributes geographical landscapes: "Capitalism strives . . . to create a social and physical landscape in its own image and requisite to its own needs at a particular point in time, only just as certainly to undermine, disrupt and even destroy that landscape at a later point in time. The inner contradictions of capitalism are expressed through the restless formation and re-formation of geographical landscapes."[4]

By combining Jameson's notion of modern feeling with Harvey's discussion of accumulation through dispossession, we can confront the ways in which green urbanism and the political practices asserted throughout this process of capital accumulation are produced by and in turn are constitutive of modern feeling and an uneven socioeconomic geography.

When we talk about green living in the city and the modern feeling it facilitates, we are not just thinking about how cities are lowering their energy consumption or the GHGs they emit or even the green spaces they have. These things are only part of the picture. I say this not to denigrate the work being done in the area of green building and green city development; to be clear from the outset, I recognize that this work is without doubt an important part of solving the climate change jigsaw puzzle. What I am interested in taking a closer look at, though, is how green urbanism and green building produce modern feeling and in turn how modern feeling overlaps the politics of climate change with the sociopolitical realities shaping cities. I propose that the process of producing modern feeling and the dynamics that it generates facilitate the production of neoliberal landscapes. This combination is placing the transformative potential of modern feeling in the service of capital accumulation. As such, in this chapter I look at how modern feeling was at work throughout the transformation of Chicago as the city moved away from its blue-collar union identity to become a global city, an image later strengthened by the brand of environmental friendliness. In conjunction, I study the socioeconomic effects of this process of transformation, all the while examining how modern feeling directs and is reproduced through the urban greening process. The guiding question in this examination is: How does the formal operation of modern green feeling align itself with a conservative impulse? And what role does capital play in this alignment?

To begin, in 2008 the world consumed approximately 8,428 million tons of oil equivalent of energy, as compared with 4,676 in 1973.[5] The world's largest GHG emitters are China and the United States. Together they contribute more than "32% of global GHG emissions, and approximately 40% of global CO_2 emissions from energy use and industrial processes."[6] Primary energy consumption is unfortunately not expected to drop or level off; the International Energy Agency predicts that, if anything, it will grow approximately 50 percent from 2005 to 2030.[7] Within this picture, the building sector has an especially significant role to play in lowering the global GHG emissions arising from global energy-consumption patterns. On an annual basis, it consumes 2,500 million tons of oil equivalent of global energy supplies. Nearly 40 percent of the world's carbon footprint comes from the built environment.[8] Lowering the GHG emissions from the building and construction industry is therefore an important factor in ameliorating climate change, and the most common ways in which this task is being broached is through policy, technological innovation, and benchmarks for green building and the development of green cities.[9]

In general, a green city has a low ecological footprint. There are a variety of ways to assess and rank a city based on how green it is. One factor is the amount of power the city receives from clean or renewable energy sources (solar, wind, hydro, and biomass) along with its overall energy use. The transportation habits of the city's residents are important indicators: traffic congestion and the number of residents who frequently walk, bicycle, carpool, or use public transportation. Data on the city's air and water quality are collected. Other factors considered are metro area sprawl, housing affordability, quality and availability of local food (farmers markets, community gardens, and markets that accept food stamps), number of areas devoted to green spaces (public parks and nature preserves), and green economic activity. Finally, a survey of the number of green buildings is conducted.

A green building is one that is environmentally friendly. The construction, performance, and demolition of a green building are designed to maximize energy efficiency and minimize environmental impacts. There are three ways in which buildings consume energy. The first is through their embodied energy—that is, the energy required to extract raw materials and to manufacture and transport the materials and components to a factory or building site as well as the actual energy used to construct the building. The second and, I should add, highest amount of energy consumed is in a building's operational phase. This is the amount of energy needed for heating, cooling, ventilation, lighting, and cooking.[10] Last, there is the amount of energy consumed at the end of a building's life cycle (demolition, landfill, recycling, and incineration).

In 2008, the U.S. building sector was responsible for nearly 40 percent of the country's primary energy consumption, which was 8 percent more than what the U.S. industrial sector consumed and 12 percent more than what the U.S. transportation sector consumed the same year. These figures are 50 percent higher than those recorded for total U.S. building energy consumption in 1980. Further, buildings account for 72 percent of overall U.S. electricity consumption. Then factor into these statistics the fact that in 2007 average per capita CO_2 emissions in the United States were 19.74 tons, as compared to 4.92 tons in China, 1.94 tons in Brazil, 1.38 tons in India, and a mere 0.64 tons in Nigeria.[11] Further, if the U.S. population was estimated at 296 million in 2005 and is projected to grow to 438 million by 2050,[12] then one pressing problem in respect to climate change is how to ensure that the new buildings needed to house this growing population might become more energy efficient.

In 1993, the U.S. Green Building Council (USGBC) was formed to respond to the negative environmental impacts of the building and construction industries.[13] In 1999, the organization piloted a digital measuring tool that could be used to assess buildings' energy efficiency. This tool is basically a rating system that in professional circles is popularly referred to as LEED, for "Leadership in Energy and Environmental Design." It has six primary categories of assessment—sustainable sites, water efficiency, energy and atmosphere, materials and resources, indoor environmental quality, and innovation—each with its own goals. One goal equals one credit point, and a total of 100 points can be given. As of 2009, the basic LEED rating requires at least 40 to 49 points; silver accreditation requires 50 to 59 points; gold 60 to 79 points; and platinum, the highest LEED rating, 80 points or more.

Michael Zaretsky points out in his assessment "LEED After Ten Years" that LEED may not be the "first green building rating system" but is "certainly the most commonly used in the U.S." Whereas the 2005 USGBC expo had less than 10,000 attendees, by 2009 this number had grown extensively, with 27,373 attendees from seventy-eight countries.[14] The growth in LEED-registered projects, membership, and professional training in the United States is testimony to LEED's growing popularity and more specifically to green living both among professionals and the general public. This is not to suggest that LEED has been immune to criticism. As many have pointed out, the rating system focuses too much on individual buildings and not enough on contextual conditions.[15] The USGBC's changing mission indicates that the organization is attempting to respond to its critics. For instance, whereas the 2005 mission statement set out to "promote the design and construction of buildings that are environmentally responsible, profitable, and healthy places to live and work," by 2010, Zaretsky explains, the organization had become more community oriented, aspiring to create, as the USGBC states, "buildings and communities" that can "regenerate and sustain the health and vitality of all life within a generation."[16]

In his evaluation of LEED, Zaretsky highlights the manner in which the rating system has "brought important attention to critical issues of 'green design,'" although he is quick to point out that there is a fundamental distinction between "green" and "sustainable" design. He calls for a "more nuanced understanding of sustainable design" because regardless of how much energy, water, or resources a building saves, we need to be mindful of the ways in which LEED facilitates the "greenwashing" of the building and construction industries. He states that the "seemingly endless number of architecturally deplorable and socially exclusive, developer-based projects" that showcase being green as a result of their LEED certification continues to raise concerns over LEED's limited scope.[17] His point, as he shows via the example of his own collaborations with a rural community in Tanzania in the design and building of an off-the-grid medical center, is that it is crucial for the integrity of sustainable design that it not be reduced to another investment opportunity. Rather, sustainable design has to aspire both to create environmentally friendly buildings and to contribute to the creation of a more inclusive society. The distinction is both a philosophical and a pragmatic one, for it concerns how we affect and are affected by the environment in which we live.

In addition to changing its mission statement, the rating system used by the USGBC's LEED program has also expanded. There are now LEED ratings for green-building design and construction, green interior design and construction, green-building operations and maintenance, green home design and construction, and green neighborhoods. The rating system for green neighborhoods specifically tackles the focus on individual buildings by providing a holistic view of the built environment. To achieve this, it combines the principles of green building and smart growth with the theories and practices of New Urbanism. The idea is to present a series of standards that can be used to create environmentally friendly neighborhoods and communities, and throughout the United States it has become a popular way to "green" cities and introduce policies that promote green urbanism. Points are awarded for green infrastructure and buildings, innovative design, smart location and linkages (site design for habitat, brownfield redevelopment, floodplain avoidance, wetland and water body conservation, to name a few), and neighborhood pattern and design (walkable streets, compact development, mixed use, mixed income, local food production, neighborhood schools, tree-lined streets, and so on). All these goals neatly align with the twenty-seven principles listed in "The Charter of the New Urbanism."[18]

New Urbanism is the outcome of a marriage between the principles of environmentalism and neotraditional design and planning. It aspires to make cities and towns more people-friendly by creating walkable, efficient, and livable communities. It is a form-generating approach to design and planning, and key components include mixed-use neighborhoods, the transformation of suburbs into communities, diversity, a celebration of local conditions, identifiable community centers and edges, infill, an integrated urban pattern, affordable housing, a variety of transportation options (public transportation, bicycle paths, sidewalks), sharing of tax revenues throughout metropolitan regions, identifiable edges and centers, interconnected streets, pedestrian friendliness, green space, diverse housing types, safe and secure environments, green building, efficient land use, and the preservation of historic buildings and sites. Narrow streets, front porches, wide sidewalks, historically inspired architectural styles, a central square, and rear garages are common neotraditional features of U.S. New Urbanism. (New Urbanism principles are also being used in Europe—for instance, in Amersfoort, Netherlands, and Hammarby Sjostad, Stockholm, just to cite a couple of examples.)

There is enormous merit to the notion that a compact, well-planned community with a strong public infrastructure and green buildings is an effective way to reduce the ecological footprint of cities and metropolitan regions. Yet as critics of New Urbanism have pointed out, if we examine some of the movement's touchstone examples—Seaside, Florida (the set for the film *The Truman Show*) and Celebration, Orlando (built by Disney Co.)—we are presented with a more ominous world of white middle-class enclaves uniformly organized by a series of restrictions and regulations. The end result is, as Michael Sorkin so scathingly describes it, a development that harbors "a single species (the white middle class) in a habitat of dulling uniformity," one that "seeks the stability of the predictable, a Prozac halcyon in which nothing can go wrong."[19]

The point Sorkin makes is important: that "New Urbanism" is a misnomer. It reproduces all the "worst aspects of Modernism" because "undergirding modernity was the fantasy of a universal architecture." Sorkin explains that New Urbanism promotes "another style of universality that is similarly overreliant on visual cues to produce social effects," wherein the uniformity of production, "the polemic of stylistic superiority, and the creepy corporatist lifestyles are scary indeed."[20] Others have argued that New Urbanism expresses a desire to return to the simplicities of traditional village life, which they have characterized as a nostalgic impulse to escape from the complexities of contemporary life. The material struggles of history are basically erased, however, and only a depoliticized historical shell is left.[21]

Although attentive to architectural history, the specificities of site, and local climatic conditions, New Urbanism surprisingly operates on the premise that there is no community prior to the design implemented by the expert knowledge of planners and architects. In order for sustainable design to be a politically transformative project, some conception of a Whole or a Universal needs to be invoked, but not in the way that New Urbanism does. A political project integrates difference and yet moves beyond identity and local differences in order to effectuate change. Unlike Sorkin, who is critical of the universal content of the New Urbanist project, I consider the problem to be less one of content than one of form: New Urbanism does not prioritize a responsive design process that emerges in collaboration with existing communities. Instead, it begins by creating a tabula rasa onto which the physical features of a traditional conception of community are rolled out. In this way, it is committed to a predefined urban form that deters the self-organization of matter and energy at work throughout the social field from challenging, disrupting, or informing the work of the designer, planner, or developer. For this reason, it is a normative approach to design that works in a top-down way to reconfigure local geographies, dehistoricizing them in the process.

Expanding on the criticism that New Urbanism shares the modernist goal of universality under a different guise—nostalgia for the picturesque small-town lifestyle—I am interested in the way that the modern feeling for green living, as epitomized by the environmentally friendly city, emerges out of an apparent contradiction as it emphasizes both the nostalgic and the new. And I have to ask whether this contradiction simply reinforces a neoliberal agenda. In addition, I would suggest that the traditional codes of New Urbanism in operation throughout the country—for example, in the greening-of-Chicago initiative—are similar to their traditional referents but serve a different function. That is, they function as a new axiom of capital.

Chicago has a long history of being one of America's main transportation hubs, which helped its economy boom as its manufacturing and retail sector grew. Yet Chicago, like other cities of the rustbelt, underwent a serious setback during 1970s deindustrialization. The latter process resulted in a rise in urban crime, middle-class flight to the suburbs, and a growth in districts maimed by abject poverty.

The social implications and causes of Chicago's demise have long been the source of great debate. For instance, sociologist William Julius Wilson argues that the poor urban black community bore the greatest brunt of the deindustrialization burden. Unemployment sent the first round of shock-waves throughout the community, and a second-round of aftershocks hit when the employed black middle class left the city. The result for poor Chicago neighborhoods, which were the subject of Wilson's study, was the corrosion of social institutions and massive social dislocation. As one woman from the South Side Chicago neighborhood exclaimed,

> It's awful the way, throughout the city, on your South and West Side, you see all these vacant lots, all these abandoned buildings, and peoples are living in the streets. Or living four and five and ten peoples in an apartment that was allocated for one or two peoples—you find eight or ten peoples because they have no place to go and no housing available. And throughout the city, you have those abandoned buildings, and vacant buildings, and just, just areas, blocks and blocks of vacant lots, where they could be building affordable, moderate-income homes.[22]

[handwritten margin note: Abandoned Buildings]

The main line of Wilson's position is that the economic downturn exacerbated preexisting social problems.

Wilson's analysis, however, sidelines the impact that postwar migration had on the social cohesiveness of Chicago's South Side, an issue Nicholas Lemann emphasizes in his epic study of the great black migration north. Lemann examines the social impact that the introduction of mechanical cotton pickers had on southern black laborers when they suddenly found their skills had become redundant.[23] For Lemann, Chicago's ghettos were an inevitable outcome of poor public policy when not enough attention was paid to job creation. He goes on to provide a brutal assessment of what he describes as a pervasive sharecropping culture—that is, the social disorganization endemic throughout African American communities, as displayed in the hustling to make ends meet, the illegitimacy, and the broken families. Last, he blames bad timing in that the migrants arrived in Chicago at a time when unskilled manufacturing jobs were on the decline.

Whereas Wilson correlates Chicago's urban deterioration with economic depression, Lehman associates the same condition with a flawed migrant culture. Both studies of race and poverty, however, contribute to a historical narrative that pathologizes Chicago's poor urban African American communities, and, as I discuss later, this patholigization reappeared in Mayor Daley's use of New Urbanism: demolishing public housing ("the

Projects") and replacing it with mixed-income housing, restructuring public schools in low-income areas, and developing initiatives to address Chicago's food deserts.

The racist view of African American communities that the 1992 U.S. public-housing revitalization program Housing Opportunities for People Everywhere (HOPE VI) was predicated on has had the unfortunate consequence of moralizing poverty. Between 1993 and 2009, a total of 249 revitalization projects were either completed, under construction, or in planning, with approximately $6 billion in grants being awarded by the program to projects across the country.[24] The program demolishes public housing and relocates residents to mixed-income developments. The idea is that by placing poor people in a context where they will have exposure to middle-class values and can interact with and make friends with the middle class, the "culture" of poverty will be broken. In reality, as Loretta Lees has amply demonstrated, the benefits of gentrification do not trickle down to the poor; they quite simply lead to "displacement, segregation, and social polarization."[25]

In an effort to attract middle-class investment in low-income areas, Chicago's low-income public-school sector was also restructured. In 2004, Chicago Public Schools launched the plan to close sixty to seventy of Chicago's low-income community public schools. Renaissance 2010, as this project was titled, would open one hundred new schools of choice: two-thirds would be nonunion, publicly funded charter schools run by the private sector; the remaining one-third would be public schools that would operate on a five-year performance contract. Pauline Lipman reported in 2009 that the plan was ahead of schedule, having closed or phased out seventy-six schools, all of which were located in Chicago's "low-income communities of color."[26]

The mixed-income, community-oriented narrative of New Urbanism not only underpins the policy to deconcentrate low-income students by sending them to mixed-income schools—operating on the premise that "advantages rest in the 'social surplus' of the middle class and whites . . . rather than in the educational resources and advantages they have accrued as a result of their status and power"[27]—but also appears in the HOPE VI program. That is, by forcing poor communities out of the "Projects," poverty will be deconcentrated, and the poor will be rehabilitated through their exposure to the more "civilized" cultural values of the middle class. The policy has proven itself to be a mask for gentrification. Lipman draws attention to the alarming statistic that in Chicago since 2000 only 1,126 of the 7,186 family units that were demolished as part of the regeneration scheme have in fact been replaced.[28]

The justification used to dismantle public-housing projects has found support among architectural and urban theorists who have been highly critical of the modernist project and more specifically of modernist high-rise public-housing schemes. The failure of modernist architecture was symbolized by the iconic demolition in 1972 of the Pruitt-Igoe public-housing project in St. Louis, Missouri. In *The New Paradigm in Architecture*, Charles Jencks comments that Pruitt-Igoe represented a "failure in planning and architecture," declaring that "modern architecture died in St. Louis, Missouri on July 15, 1972 at 3:32pm (or thereabouts) when the infamous Pruitt-Igoe scheme . . . [was] given the final *coup de grace* by dynamite."[29] One of the most influential criticisms of modernism had come from Jane Jacobs many years earlier, who influenced New Urbanism with her 1961 groundbreaking book *The Death and Life of Great American Cities*.[30] In it, she put forward a convincing case for community-oriented urban planning, arguing that the crime and social alienation characteristic of top-down modernist planning and the high rise had weakened urban life. To counter these problems, she recommended efforts to revitalize the local economy with more diversity and greater density. Out of this New Urbanist trajectory, mixed-use and mixed-income developments were held up as the new design idiom that could replace the isolation and social segregation of the high-rise projects.

The pathological codification of disenfranchised communities and the spaces they inhabit has given rise to a series of public-policy initiatives that have reconfigured the geography of race and class in many U.S. cities, and Chicago has been no exception to this trend. For instance, policies of deconcentration have been part of a larger effort on behalf of Chicago's governing authorities and corporate sector to pull Chicago up by the boot-straps and transform it into a "global city" so that its elite can competitively position itself in the global free market.

"Globalization is a trillion puppeteers dancing on very long strings," explains Saskia Sassen, and "global cites are where the puppeteers live and work."[31] Chicago has grown to become a place of "power and sophistication," says Adele Simmons: it boasts having a multicultural base, with 22 percent of its population being immigrants who send approximately $1.8 billion dollars home each year; it has 130 non-English newspapers to meet

Chicago Changing

Creative Class

the needs of its cosmopolitan population; and it is home to seventy consular offices to meet the demand that comes from being a truly "global" entity. The turnaround came about largely because the city was flexible and open to new ideas, which translates into being willing to diversify its economy. And the prize: "Being known as a global city is an economic and cultural *asset*"![32] As a result, Chicago became a destination for corporate investment and relocation, one famous example being Boeing Company, which relocated its headquarters to Chicago from Seattle in 2001.

Much of Chicago's success comes from being able to lure the creative class, those well-educated individuals who flock to a city proudly showcasing its diverse cultural assets, its wide array of recreational options (restaurants, cafes, bike trails, water sports, etc.), an efficient transportation network, a variety of upscale condos, employment in the knowledge economy and intellectual service industry, along with opportunities to work in elegantly designed offices in the city's downtown. What is more, today Chicago has the added benefit of enjoying the status of being a wonderful example of sustainable urban living: compact and dense, it boasts a strong public transit system; it is walkable and bikeable; and its residents enjoy energy-efficient buildings, mixed-use developments, and green spaces.[33] It has basically transformed itself into an urban success story.

Popular culture presents a consistent image of what a successful city looks like. This city features in advertisements, provides the backdrop to television shows and movies, and is marketed as a tourist attraction. The successful city is fashionable, fast paced, sexy, wealthy, globally linked; is able to attract upwardly mobile immigrants and swarms of visitors; and, more important in the twenty-first century, is green, clean, and diverse. As stated in the New Urbanism news publication *New Urban Network*, "The Daley administration and other public entities have accomplished many positive things—among them, demolishing unlivable public housing towers at Cabrini-Green; sprucing up city parks; rehabilitating architecturally distinguished old schools rather than building second-rate replacements; improving the public transportation system; and ending the expedient practice of putting new schools on park land."[34]

Cities account for up to 70 percent of GHG emissions, and yet they cover only a mere 2 percent of global land area. The majority of the world's population resides in cities, with many more expected to follow in the footsteps of the millions of people worldwide who have already migrated from rural areas.[35] For these reasons, there is enormous potential to lower global

GHG emissions by making cities more environmentally friendly. So when Mayor Daley publicly kicked off his goal of making Chicago the greenest city in the United States by symbolically installing a green roof on City Hall in 2000, he positioned himself at the cutting edge of environmental urban politics. In 2006, he went on to adopt the Environmental Action Agenda as a way to pragmatically green the city's infrastructure, buildings, and energy as well as to improve livability standards for Chicago residents. The list of green urban achievements that took place under Daley's watch is indeed impressive: 1,300 new acres of open space, 7 million square feet of green roofs (which reduce urban heat island effect and help manage storm water), 90 miles of green street medians installed, half a million trees planted, removal of one of the downtown airports for a 100-acre park instead, 165 miles of bikeways built, and new investment pumped into the city's transportation infrastructure.

Then on September 18, 2008, Mayor Daley announced his Chicago Climate Action Plan. The plan set out to take a serious stab at reducing the city's carbon emissions. It spearheaded the city's energy use and aimed to reduce waste and finance emissions-reduction programs.[36] Overall, the aim was to place Chicago at the forefront of environmentally friendly urban planning. Some of the strategies implemented include changing building codes and retrofitting residential, commercial, and industrial buildings so as to reduce the energy they use and improve their water efficiency. New policies require all new public-building renovations to comply with green standards. Some of the mitigation strategies for clean and renewable energy sources outlined in the plan include upgrading or repowering the twenty-one Illinois coal plants that supply energy to Chicago, improving power-plant efficiency, and procuring renewable energy. The question that remains is: How does the green glow of the Chicago brand infuse everyday life on the ground?

The 2009 U.S. Census Bureau American Community Survey reported that Chicago continues to suffer from severe income disparities, with the average African American person earning $28,725 a year, but the average white Chicagoan earning $63,625 a year.[37] The only other large U.S. city that topped Chicago for inequitable income distribution across racial lines was Dallas. In 2009, Chicago's rate of violent crime (murder, rape, robbery, aggravated assault, and aggravated battery) was approximately double the national average.[38] These statistics point to disparities in income and opportunity that the greening-of-Chicago mission has certainly not addressed.

Chicago may have become more environmentally friendly, but if we look at the growing wage inequality that has occurred despite "robust job growth at the bottom of the labor market" in Chicago,[39] the neoliberal transformation of public housing, and the restructuring of the public-school sector, in particular those schools that serve low-income communities, then the city as a whole remains an unfriendly place for its poor and nonwhite residents.

In light of this assessment, I suggest that the modern feeling for New Urbanism and the green city in operation throughout Chicago does not change the axiomatic of capital; it merely modifies how that axiomatic works. Deeply critical of suburban sprawl, New Urbanism advocates for the importance of public spaces, transit-oriented development, density, and community vitality. Thus, in principle it appears to offer an alternative to the neoliberal focus on individualism, privatization, and the laissez-faire approach to urban design that has resulted in suburban sprawl, an automobile-centric culture, and the reification of private space. As the alienation of a capitalist social order has dismantled the vitality of community, a problem has arisen in which the reinvestment of the energies of alienation in a nostalgic view of a traditional social form is deployed as a signifier of the New.

The problem I am outlining here is formal; it concerns the manner in which the content of the past is used to accomplish a neoliberal agenda in the present under the guise of fulfilling a transformative program. The green roof on Chicago's City Hall is just another code, alongside other codes such as the LEED-rated buildings, housing voucher schemes, bicycle paths, and so on and so forth. What grounds all of these codes and the shifts they undergo over time is the axiomatic of capital, for in all cases capital serves as the justification for urban development and change. As Gilles Deleuze and Félix Guattari have so aptly noted, "The strength of capitalism indeed resides in the fact that its axiomatic is never saturated, that it is always capable of adding a new axiom to the previous ones."[40] How capitalism overcomes contradiction is by introducing a new axiomatic— the global city, the environmentally friendly city. None of these new names, however, actually changes the axiomatic of capital; they simply obscure how it works.

By turning green urbanism into an economic function, Mayor Daley integrates the Climate Action Plan and the urban greening strategies into the axiomatic of capital. Presenting green urban development, investment, and spending as a way to solve the financial crisis establishes an ideological opposition between the good green successful city and its urban Others (at the regional scale, the shrinking cities of the Rustbelt; at the metropolitan scale, the foreclosed landscapes of the suburbs; and in the urban core, Chicago's poor communities' living in the Projects). The environmentally friendly city is just another way to organize everyday urban life around consumption—a way that helps promote Chicago as a tourist and convention destination.

There is nothing particularly new in claiming that there exists a strong connection between capital accumulation and urban transformation. David Harvey is one notable theorist who has studied this phenomenon at length. Leaning on Marx, he outlines that capital transforms the conditions of production to procure a surplus-value and accumulate further capital. In *A Brief History of Neoliberalism*,[41] he succinctly defines neoliberalism as a theoretical position that valorizes individual freedom and liberty, which in practice is propped up by a system of deregulation, free markets, and private-property ownership. He describes the ways in which neoliberal theory and practice have resulted in the withdrawal of government from everyday life, such that the role of the state has changed from facilitating social well-being to becoming largely concerned with upholding the interests of the free market and safeguarding private-property rights.

Economic power and influence is enhanced through what Harvey describes as a relentless process of accumulation by dispossession: individuals, communities, and entire societies are dispossessed of their assets, resources, wealth, and rights. For example, in Chicago the Projects were developed into mixed-income communities, and vouchers were provided to those who could not return to these communities that they could use on the private rental market; and public schools in low-income neighborhoods were restructured to make these areas more attractive for middle-class families. Throughout all these measures, however, the voices of the poor were rendered inaudible.

Another feature of the gentrification process has been the concerted effort to address the perennial problem of food deserts throughout Chicago. People living in these areas that lack access to grocery stores with healthy food options experience on average greater diet-related health problems (obesity, diabetes). In the United States, food deserts share the following demographic characteristics: low income, poor education, and high populations of African Americans. In 2009, Mari Gallagher Research and Consulting Group reported that since 2006 "Chicago's food desert shrank by 1.4

food deserts

square miles, 220 blocks, 4 Census tracts and roughly 24,000 people," Yet, despite this change, the desert "remains large, affecting 609,034 residents as of September 2008." More important, the "majority of people living in food deserts continue to be African Americans by a wide margin," of which nearly "200,000 of them are children and more than 100,000 are single mothers."[42] Despite the improving access to healthy affordable food in Chicago's food deserts, there still exists a racist, classist, gendered divide regarding who benefits from these greening initiatives. The point in all this is that the gentrification process does not signal greater equality.

Capital accumulation shapes the social organization and built environment of the city through policy, planning, and development, all of which become defined by capital as another form of productive labor. As a key ingredient to the production of surplus value, the environmentally friendly urban landscape basically functions to keep capital in circulation. It is no coincidence that in 2003 Mayor Daley raised 25 percent of his campaign contributions from developers and real estate (Dick Simpson and Tom Kelly report that 27 percent came from financial services and law firms, 11 percent from wealthy individuals, and only 10 percent from labor unions).[43] Meanwhile, Renaissance 2010 was first proposed in 2003 by the Commercial Club of Chicago, an organization that Lipman explains is made up of Chicago's most "powerful corporate and financial leaders." The organization's goal is to end what it perceives to be the "monopoly of public education."[44] Transforming the public schools of low-income areas is an integral ingredient to Mayor Daley's gentrification of Chicago because middle-class people are attracted to areas that have good schools. Interestingly, the Renaissance 2010 schools were chosen by Chicago Public Schools and the Renaissance School Fund, an organization established by the Commercial Club of Chicago.

If we expand Marx's notion of value (use-value, exchange-value, surplus-value) to posit that modern feeling is a value that in the final instance exceeds value, in this context modern feeling is both abstract and concrete labor. It is worthwhile adding here that I am the first to admit this is a rather idiosyncratic use of Marx, who in volume 1 of *Capital* clarified the distinction in these much cited lines: "On the one hand, all labour is an expenditure of human labour-power, in the physiological sense, and it is in this quality of being equal, or abstract, human labour that it forms the value of commodities. On the other hand, all labour is an expenditure of human labour-power in a particular form and with a definite aim, and it is in this quality of being concrete useful labour that it produces use-values."[45]

For Marx, the concrete labor of the individual laborer selling his or her labor as a commodity works to produce use-value (commodities have a use-value and an exchange-value). At the same time as the laborer sells his or her labor, abstract labor enters the equation as a social relation (commodified labor power that is exchanged and is integral to the circulation of capital).

The concrete labor expenditure of modern feeling—the mayor working to rebrand the city, individual laborers tearing down the Projects and replacing them with mixed-income developments or working on the reorganization of the city's fabric through green infilling, and so on—can be said to acquire a social quality when these individual projects combine to rebrand the city anew. The new Chicago brand is social, and it is behind the city's rising success. For instance, in 2009 Chicago hosted 31.9 million visitors, of which 28.9 were domestic travelers and 11.7 million domestic business travelers, contributing more than $10.2 billion to Chicago's economy.[46] In 2009, approximately 1,130,000 overseas travelers visited Chicago, ranking it tenth for visitors in the United States.[47]

The logic of the commodity is central to urban modern green feeling because modern feeling works both as a use-value (decoding the urbanscape, reconfiguring urban geography, and reorganizing social life) and as an exchange-value (climate change politics and urban rebranding). Further, the structure of modern feeling (the new, erasure, etc.) is also the outcome of "value in motion," as Harvey puts it.[48] Harvey fittingly insists that before a built environment is even produced, developers and entrepreneurs have already invested their capital. As a spatially fixed commodity, real estate has to be produced first and later sold. The delay between investment, production, and exchange presents a risk because circumstances can dramatically change from one stage to the next (the recent global financial crisis has aptly demonstrated this point). The state responds by freeing up the spatial fixity of real estate—Mayor Daley's demolition of the Projects and the subsequent transformation of them into "mixed-income" properties; the racial and income inconsistencies of food-desert initiatives; and the mixed-income public schools—to assist the flow of capital. Modern feeling is appealed to in order to legitimate rapid and autocratic measures, and it functions as a structure that creates new openings through which capital flows can pass. The social inequities and hierarchical arrangements defining the geography of Chicago intermingle with urban policies and developments whose primary aim is to attract investment and make possible capital accumulation.

The phenomena of modern feeling liberates the old dichotomy that held the urban core in opposition to metropolitan sprawl, recoding the urban landscape as it thrusts it into a different historical situation (green urbanism, climate change, neoliberal urban planning and policy), such that the content of urban life is reduced to an exercise in branding. Through modern feeling, capital deterritorializes the city—from manufacturing city to global city and then later to environmentally friendly city—and along with each renewal the city as a lucrative form of production is revitalized, and urban geography is reterritorialized by capital again. What needs to be mentioned, though, is that the "city" is not a unified homogenous entity; some urban spatiotemporalities facilitate the flows of capital, but others (low-income public schools, the Projects, and so on) block its movement. As a result, these shifting conduits of capital, goods, people, real estate, commerce, and ecological forces differentiate the city's spatial dynamics.

How does having the identity of a modern urban dweller, a person living in a green home in a green city, make us *feel* about ourselves? The question that remains is: What kind of modern feeling is this situation producing? Although the retrofitting, infilling, and new green construction, the mixed-use development, and the improvement in local food sources are laudable initiatives in their own right (they provide important experiments in lowering the GHGs emitted from the built environment), we also need to remain alert to the principles of neoliberalism accompanying this trend. The greening of Chicago might have made the city a more environmentally friendly city for residents and tourists alike; it may have helped turn Chicago into a destination spot, attracted outside investment, and promoted an extensive program of urban redevelopment. However, the benefits arising out of Chicago's greening process have not been equitably distributed.

This inequity raises an interesting point concerning community and failure, both of which I treat as analytical categories. A community is not a site; it is a spatiotemporal event, the dynamics of which cannot be flattened in the way New Urbanism tries to do. As depressed parts of the city are transformed, community participation is a crucial ingredient in creating appropriate solutions specific to the individual and collective needs of any given community. Furthermore, my intention in pointing out the "failures" of Chicago's redevelopment and transformation is not to suggest that Chicago itself represents a failure of contemporary urban green design (such a claim would simply repeat the modern feeling implicit within the failures described thus far). As a way to challenge the conviction that nothing can ever be the same again, failure needs to be infused with a utopian thinking that sets out to self-reflexively engage with the past and quite simply try again. But this process also demands an attitude of critical realism that is historically informed and situated in order to remain alert to the ways in which the axiom of capital functions.

There is an exteriority to the principle of sustainability that New Urbanism and green design just do not share. One of the key principles guiding the various movements of sustainability is a commitment to connect environmental and social justice concerns, all the while retaining the exteriority of the one from the other. In contrast, the greening of Chicago initiative conjoins the two. Whereas the principle of sustainability seeks to expunge the notion of the Other as it is represented throughout the politics of climate change, green urbanism, the removal of urban "blight," and urban revitalization, the Chicago green initiative depends on the dualistic structure inherent in these representations (positive/negative, good/bad, prosperous/deteriorating). The answer to climate change and the faltering economy of Chicago is therefore nothing new if what we understand by the "new" engages a Marxist problematic of how effectively to challenge neoliberal capitalism and the concomitant problems of privatization, free-market economics, financialization, and commodification.

8

SPILL, BABY, SPILL

The chant is "Drill, baby, drill."

—Alaska governor Sarah Palin at a vice presidential debate, October 2, 2008[1]

Sub-Saharan Africa is likely to become as important a source of U.S. energy imports as the Middle East.

—Anthony Lake et al., *More Than Humanitarianism*[2]

I ask the Congress to commit $15 billion over the next five years, to turn the tide against AIDS in the most afflicted nations of Africa and the Caribbean.

—President George W. Bush, 2003[3]

In what seemed like a cruel twist of irony, the *Deepwater Horizon* oil rig sank on Earth Day 2010 after exploding two days earlier, on April 20. The rig and blowout preventer was owned by Transocean and leased by BP, the largest oil and gas producer in the United States.[4] The spill was the largest accidental oil spill in history. In the aftermath of the spill, locals set up a sign in Belle Chasse, a fishing village in southern Louisiana, that read: "Damn BP! God Bless America!" As the sign indicated, the consensus was that BP was to blame, and in more senses than one it was. But as the finger of blame pointed in the direction of the oil mogul, one might have felt uneasy with how quickly and easily the United States, the largest oil-consuming nation in the world, blamed the foreign oil company. It all seemed too straightforward, too unified, and too uncontroversial for what was such a politically divisive situation. What did this supposedly uncontentious finger of blame conceal from view? The short answer: the violence of oil capitalism. I use the term *oil capitalism* purposefully to highlight the confluence of unregulated free markets, militarism, and environmental degradation.

The politics and management of oil resources and revenues at the center of global oil operations is well documented, as is the complicated connection between violence and oil.[5] Disturbing images of the *Deepwater Horizon* drilling rig exploding and the aftermath of oil plumes, tar balls, toxic sludge, and wildlife drenched in molasses-thick oil were widely disseminated throughout mainstream media. The mobile rig was situated approximately forty-one miles from the Louisiana coast, and it burst into flames when methane moved up the drilling column, killing eleven platform workers and injuring seventeen more. When two days later the rig sank, oil from the broken wellhead gushed uncontrollably into the ocean. In response, the U.S. public was outraged by what they witnessed. Yet the same public is far less damning of the oil industry and government when faced with other oil-related images of cars in traffic jams or smog on the horizon. It feels no outrage as it consumes the largest amount of oil in the world and relies on oil imports in order to do so. And its anger was indeed short-lived. Within a year, the U.S. public was once again supporting energy exploratory initiatives even if they were made at the expense of environmental concerns. So why is it that the images of the *Deepwater Horizon* oil spill sparked such a vacuous public outcry? It seemed more as if public outrage obfuscated the real source of violence endemic to fossil fuel production and dependency.

The October 2010 issue of *National Geographic* featured images of bottlenose dolphins sliding through the oiled-slicked waters of Chandeleur Sound, Louisiana; a dead turtle stranded in a reddish brown bed of sludge in Barataria Bay; and workers' bodies smudged brown as they bagged oil throughout the Louisiana wetlands. For the three-month period following the spill, the *New York Times* regularly showed updates, such as images of the *Deepwater Horizon* rig collapsing amidst an orange glow of raging fire and aerial views of the oil-drenched waters of the Gulf of Mexico. *CBS News* presented its viewers with footage of wildlife and clean-up workers dripping in oil. The English newspaper *Telegraph* displayed images of oil encroaching on shoreline communities, and the *Guardian* showed pictures of oil slicks producing rainbow patterns over the surface of the ocean. Reuters documented workers dressed in protection suits collecting oil-absorbent, water-repellent booms soaked with oil and the spill surrounding the marshland south of Venice, Louisiana. And similar images appeared in the global media. This provocative visual landscape interrupted everyday life the world over, especially in the United States.

The immediate political fallout of the spill was the retraction by forty-fourth U.S. president Barack Obama of his plans to end a moratorium on oil exploration that he had only recently publicly announced (March 2010).[6] Prior to the *Deepwater Horizon* spill, the president had been preparing to move ahead with a contentious plan to open the Atlantic coastline, the eastern Gulf of Mexico, and the northern coast of Alaska to offshore drilling for oil and gas. Like his presidential predecessors, Obama considered oil a strategic resource. He hoped the policy would win him much needed political support for his energy and climate bill, produce revenue from selling offshore leases, and reduce the U.S. dependence on oil imports. The catastrophic spill in the gulf changed all that, and in May 2010 the president announced a six-month moratorium on deep-water (more than five hundred feet) oil and gas drilling. A political domino effect was in play as Vermont senator Bernie Sanders (I) proposed legislation prohibiting offshore exploration and drilling throughout the Atlantic and Pacific coasts as well as a large part of the gulf and that fuel-mileage standards for automobiles be increased to fifty miles per gallon.

On the ground, U.S. public opinion toward energy exploration and environmental protection dramatically changed direction. A May 24–25, 2010, *USA Today*/Gallup Poll reported that Americans had started to prioritize the environment over the development of energy supplies. The report summarized its findings thus: "In March, by 50% to 43%, Americans said it was more important to develop U.S. energy supplies than to protect the environment, continuing a trend in the direction of energy production seen since 2007. Now, the majority favor[s] environmental protection, by 55% to 39%—the second-largest percentage (behind the 58% in 2007) favoring the environment in the 10-year history of the question."[7]

As emotive imagery of the *Deepwater Horizon* event flooded the global media for eighty-eight days, the public watched the irruptive and unpredictable forces of material life in violent confrontation with human hubris. The shock brought on by the apocalyptic visual landscape defied interpretation. Indeed, for a brief moment, no matter how widely the event was reported, it could not be easily inscribed with meaning. The images were so relentless and the situation documented so uncompromising that public debate over the future of the world's energy and over humanity's unwavering dependence on fossil fuels was sparked. Media coverage of the event prompted momentary wrinkles to form on the face of climate change discourse in large part because the U.S. public could no longer look the other way. The everyday denial that allows people to continue with business as usual, the Freudian disavowal of "I know, but . . . ," was no longer effective.

One explanation for the sharp change in U.S. opinion comes from cultural anomaly theory. The shock of ecological disaster plays a transformative role in society, for it represents "an anomaly in the institutional order"—an anomaly that signals the "fundamental challenge" "to actors' identities within an existing institutional order."[8] Cultural anomalies redefine social problems by focusing public attention on a specific issue. The approach adopted by cultural anomalists is intriguing, for they suggest that an otherwise incomprehensible situation (that is, the basis of a shock) is able to "represent" an interruption at an institutional level. But isn't a trauma traumatic *because* it defies representation? That is the whole point of going to therapy to deal with trauma. In therapy, cure comes from narrating the sequence of events that make up the history of a trauma, and in so doing the shockwaves trauma produces are supposed to be neutralized. The premise of trauma therapy is basically that the violence of shock induced by trauma can be successfully defused when it enters language and is reconstructed within a symbolic field.[9]

Perhaps the anomaly that ecological disaster images present comes from the way in which they cannot be rendered meaningful. Where the politics of such shocks lie has less to do with what they represent and more to do with the manner in which they defy representation. In fact, I argue that it is when the incomprehensibility of a shock is represented that it becomes coherent and its affective power is depoliticized. In other words, I would like to depersonalize cultural anomaly theory. Where cultural anomaly theory wonders why the "linkage of the spill with other problems and issues in the oil industry by institutional entrepreneurs has so far failed,"[10] I would add that it is too easy to focus on the failure of institutional entrepreneurs and that we need to take a hard-nosed look at how we facilitate this failure. Environmentalist Bill McKibben aptly explained for *U.S. News & World Report*, "This is one of the moments when we're offered an opportunity to really see what's going on in the world. . . . Even if that oil made it safely to shore and got burned in the gas tanks of our cars, it would be an environmental catastrophe."[11] In other words, those of us who live high-oil-consuming lifestyles need to stop pointing the finger of blame in the other direction and start pointing it toward ourselves.

And in terms of what slipped through the cracks of signification, the *Deepwater Horizon* oil spill presented the disruptive brutality of material life, thereby posing an outside to thought.[12] The incomprehensibility of this outside occupies an important formal position. In its alterity, it poses a dramatically different situation to the one we currently find ourselves in. William Freudenburg and Robert Gramling write in *Blowout in the Gulf*:

> The explosion of the *Deepwater Horizon* provides, in the most vivid form that any of us would ever want not to see, not just a tragedy, but also a challenge, and an opportunity—a challenge to take a closer, more clear-eyed look at our policies, and an opportunity to realize that this is a hole that cannot be escaped simply by digging deeper to look for more oil. Instead, our only hope for a better energy future is to respond to the oil-darkened waters with clearer thinking—to move now to confront the reality of using ever-increasing quantities of scarce and precious petroleum, and to begin the move to a future that will be controlled by our decisions, not by our dependence on the fast-disappearing remnants of the time when dinosaurs last roamed the earth, a good hundred million years ago.[13]

The key here, as Freudenberg and Gramling so fittingly note, is that the explosion presented what had previously been unseen and unheard. The blowout in the gulf was almost like a repressed unconscious spewing forth into consciousness; it was something that "any of us would ever want not to see." The crucial point I take away from this comment is that people do not want to see how they are implicated in the very policies and lifestyles that produce violence of this sort—much like the sign that read "Damn BP! God Bless America!" Although the Louisiana locals' anger is understandable, those who posted the sign at the roadside need to be reminded that it is not just any policy but "our" policies (theirs included) that their sign conceals from view. Although it is important that those directly responsible are held accountable and are forced to incur the costs of cleanup and recovery, instead of just focusing on who is to blame, how might we understand all the failures that led to the spill as part of a larger narrative?[14]

As oil soaked the ecosystems of the wetlands, the Gulf of Mexico, wildlife, and local economies, the outrage experienced by the U.S. public in response to the spill produced a kind of ideological interpretation of the event—"BP is to blame." I do not mean to suggest that BP is immune from being held accountable and responsible for its operations in the Gulf of Mexico that produced the spill. I do argue, however, that blame inscribed the event with moral meaning and in so doing defused the political potential of the shock.

McKibben hit the nail on the head when he declared:

> Stop pretending that the fight is over energy independence or oil security. We need to tell the truth. The pollution you can see, like the spill in the Gulf, is the least of our problems. What stalks our future is the invisible damage done when the structure of the CO_2 molecule traps heat that would otherwise radiate out to space. It's not when BP makes an outlandish mistake; it's when BP and Exxon and the rest of the fossil fuel industry carry out their daily business. It's not when things turn black; it's when they turn hot.[15]

It is not the allocation of blame that is at fault, but rather what this allocation of blame produces: denial. The real issue at stake is that the world's climate is changing, and there is no difference between the pollution emitted from cars every year and the pollution that comes from oil spewing into the gulf. In this regard, the significance of failure does not lie in the content of historical failures, but rather in the formal position such failure occupies and subsequently the kind of being constituted by this position. Failure can also be a source of liberation, but only if it is released from the shackles of negativity and is allowed to activate a sense of optimism for the future as a source of change for the present: failure as an opportunity for liberating change.

This book has centered on this very issue: given the scientific consensus over the irreversible and harmful social and environmental consequences that a few climatic changes in degree will cause, it is vital that business as usual be disrupted. Enough with the disavowal, the "damn BP," "damn the corporation," and so on! Damn us all for damning everything but ourselves—the citizens of high-consuming societies who are implicated in the endless production and consumption of material life.

With high-oil-consuming populations working diligently at being more ecologically aware, recycling waste, eating vegan, wearing organic cotton, offsetting carbon footprints, driving hybrid automobiles, and so on and so forth, it is important that one does not lose sight of the larger picture of how capital accumulates and at whose expense. As capital accumulates, so too do injustices and inequities the world over. Simply modifying the current system is therefore not enough. President Obama's six-month moratorium is an obvious modification, but so too is a permanent ban on offshore drilling and exploration. I can already hear my critics scream: "So you are damned if you do and damned if you don't!" Not quite: I am suggesting that much more is needed if we are going to stop our dependence on fossil fuels and the structures of violence from which that dependence arises.

Doing something transformative in response to climate change, environmental exploitation, the degradation accompanying poverty, and the myriad ways in which these issues intersect is going to require the coordinated effort of the international community working in collaboration with local communities and regional governments to ensure that as new measures are instituted, the benefits accrued are equitably distributed and that public institutions are established to oversee this equitable distribution. This effort obviously needs to be made with more urgency. At the same time, "urgency" cannot be viewed within a neoliberal lens. It cannot provide yet another "crisis" modality through which draconian and conservative restructuring measures are enforced.

As we face the *crisis* of climate change and our dependence on fossil fuels, it is important that we put to work Naomi Klein's explosive and bold thesis of crisis capitalism.[16] We cannot permit the shockwaves of environmental and climate crises to be exploited as another excuse to impose free-market policies and practices. In other words, we cannot afford to succumb to the shock therapy imposed under the different guises of disaster capitalism described throughout this book: neoliberal capitalism, climate capitalism, environmental capitalism, and oil capitalism (all of which, I should add, are implicated in one another). More critical realism is needed, and, as horrible as it may sound, we need first to give up on the idea of "nature."

In the context of pollution, starvation, species extinction, toxicity, contamination, and the rubble left behind by warfare and ecological disasters, mythologizing the natural world as a pristine untainted virgin space—the antithesis of the "artificiality" posed by technology and industrialization—is quite simply a waste of energy and at worst a displacement activity. Contemporary life is a cocktail of material life, technology, human activities, and the reproduction of capital. And the cocktail has to be one that people are more willing to imbibe if equitable pragmatic responses to climate change and the social inequities it will precipitate are to be implemented. This cocktail has a name: machinic life.

Machinic life is not the same as material life because the affective combination of energy, matter, force, surplus-value, and machines may be composed of material life but not be reducible to it. Machinic life confounds the clear-cut boundary between the artificial and the natural. It renders futile the impulse to delineate between the natural and the artificial. Trying to ascertain where the natural ends and the artificial begins is quite simply useless. After centuries of industrialization, what is called "natural" can only be an expression of a deep-seated anxiety and hostility toward change. Introducing critical realism into the current situation means posing the following two questions that machinic life presents: the aesthetic question "What kinds of affective and sensorial organizations does machinic life produce?" and the political problem "What motivates how machinic life is used and the subsequent organizations this use creates?" And here we can return to the images of the *Deepwater Horizon* spill.

The *Deepwater Horizon* oil spill raised questions over the future of energy, offshore drilling, and stricter oversight of the oil and gas industries. The ecological catastrophe quite literally shook the public imagination, forcing into view the violence of humanity's dependence on fossil fuels.[17]

The disaster imagery of the *Deepwater Horizon* bore witness to a struggle that took place between the subject and object of images and the affective landscapes coordinating these images; between the act of negation (as the workers attempted to impose a purposive form on material life in tirelessly trying to cap the well) and the affirmative power of creative production (as the well continued to spew forth oil for eighty-eight days despite professional experts' efforts to cap it). The images of ecological disaster that appeared in the immediate aftermath of the explosion defeated the omnipotence of a rational thinking subject—a view that presupposes that subjectivity is basically immaterial. No amount of reasoning seemed to be able to bring the situation under control: contractors tried to cover the wellhead with a dome and then to close it with a submersible robot; then they frantically used chemical dispersants to break up the oil and eventually scooped up the gunk by hand, until finally a solution was found: the well was successfully capped and then filled with cement on July 15, 2010. The well was officially declared dead on September 19, 2010, five months after the explosion.

Despite the U.S. Oil Pollution Act of 1990 (Pub. L. 101-380),[18] no amount of money spent by BP or military resources supplied by the U.S. administration could sufficiently contain the spill before it wreaked havoc on wildlife, ecosystems, and local communities. The crisis dramatically worsened before it could be resolved. Out of this animated combination of forces, energies, and affects, the aesthetic figure of machinic life presented a violent image of mythic proportions: blue skies were interrupted by clouds of thick black smoke; the ocean was ablaze; and pelicans were either freezing to death or frying alive because oil had destroyed their feathers' insulation capabilities.

Footage of the oil spill was quickly disseminated across the globe on television media and other news outlets, prompting public outrage over the crisis. With this outcry came a momentary shift in attitude. Confronted with an ecological catastrophe of such magnitude, people could no longer evade the violence of machinic life with which their everyday lives are complicit. This confrontation radicalizes Marx, who so astutely observed that through labor human beings create their means of subsistence by imposing an objective form onto material life, thereby changing it into a commodity and subjecting it to use- and exchange-value. Through this process of alienation, capital is set in motion. Yet despite humanity's efforts to objectify material life, it indefatigably asserts its autonomy by challenging both capitalist modes of production on its own terms and humanity's conservative belief in "nature" as an organic balanced Whole that we exploit and subsequently upset the equilibrium of.[19]

The imagery of ecological catastrophe points to yet another dynamic at work in humanity's relationship to material life, one that was not the focus of Marx's thinking: the connections human beings establish with material life are not just negative; they are affirmative, creative, and sensuous. In many respects, machinic life is the effect of the intimate connections people establish with technology, material life, and capital accumulation. Just think of all the information people willingly offer up about themselves to Facebook, eBay, online banking, and so forth or of the implants that allow the deaf to hear again or of the pacemaker that improves debilitating heart conditions—all these things are examples of intimate technologies. The point is that bodies, technologies, material life, and capital are implicated in each other but cannot be reductively equated with one or the other; they resonate, activate, and energize each other. The affective processes coordinating machinic life means aesthetics is well positioned to pose new ways of engaging with the effects of machinic life. Through the politics of aesthetics, people are not reduced simply to making one another and the world in which they live, as Marx proposed; rather, to borrow a concept from Jacques Rancière, through the redistribution of the sensible people are better positioned to remake themselves and the connections they form with other entities and with the environments in which they live.[20] The key question is: How can people start loving machinic life for what it is, embracing it in all its brutality and sensuality?

As we witnessed the waters of the Gulf of Mexico uncontrollably burst into flames as the blowout in the 18,000-foot Maconda well refused to be tamed, a horrifying materialism presented itself. This is important because as an "outside," machinic life seemed to keep humanity on its toes, defiantly gushing approximately 56,000 to 68,000 million barrels of oil into the ocean daily for a total of eighty-eight days, regardless of how many people worked around the clock trying to contain the spill.[21] The approximately 4.4 million barrels of oil that filled the ocean during that time brought the oil mogul BP to its knees, savagely destroying the already weak economy of the gulf and leaving wildlife and wetlands in the region saturated in black gunk. In other words, through its radical alterity, machinic life intruded upon humanity's sense of self-importance and the hubris underscoring this sense, for the miscreant Maconda well proved very resistant indeed.

The spectacle of ecological disaster such as the destruction left behind by the oil spill in the gulf subjectivizes violence. Charged with a sensorial power, the media images documenting such events pulsate across the globe. Yet even though such images might clearly identify the victims of violence (people, ecosystems, cities, wildlife) and the actor of violence (machinic life), they conceal what Slavoj Žižek calls "objective violence." As he says, objective violence is "violence inherent in a system: not only direct physical violence, but also the more subtle forms of coercion that sustain relations of domination and exploitation, including the threat of violence."[22] Objective violence operates at a structural level, allowing business to carry on as usual.

The objective violence that enables the average person living in an affluent country to lead a reasonably comfortable life is made possible by the myriad ways in which the violence of oil production and the geopolitics of oil operates throughout the world and has done so throughout history. For instance, oil impedes democracy by making governments of poor states who export oil more authoritarian. The first reason for this outcome is the rentier effect. The term *rentier* refers to states that obtain a large portion of their revenues from external rents—namely, foreign states, interests, or individuals.[23] Oil money is used to placate the population against holding the government accountable for what they suffer, resulting in the government's being less likely to subsidize development programs.[24] This placation then feeds into a repression effect as the government uses its fiscal power to dull popular opposition, in turn producing more authoritarian forms of governance. Third, as there is less investment in the social field (education, health, services, infrastructure), the population is less likely to push for democracy.[25]

The civil war in Sudan, the largest country in Africa, is a good example of how the subjective violence of the *Deepwater Horizon* event is undergirded by systemic forms of violence central to the geopolitics of oil. The inequitable distribution of the costs and benefits of oil production fueled the Sudanese civil war. The war resulted in approximately 2 million deaths, 4 million internally displaced, 420,000 refugees, and approximately 2,500 rebel child soldiers.[26] The U.S. Committee for Refugees and Immigrants reported in 2001 that oil revenues had provided the Sudanese government with "substantial new revenue that enabled it to double its military expenditures compared to 1998."[27] When the Sudanese government in the North appropriated oil-bearing lands in the South, it expelled the indigenous population from their lands,[28] which incited violent protests by locals who depended on the land for their livelihood and the subsequent creation of the Sudan People's Liberation Army (established in 1983 by John Garang, a senior army official who defected from government forces).[29] Chevron Corporation (the first oil company to establish a permanent facility in Sudan), which had supported the Numeiri government of the North and later the Baggara militia in return for secure access to the oil-rich regions of the South, began withdrawing from the region in 1984 after its operations were attacked and three Chevron employees were killed by the Liberation Army.

Similarly, in Nigeria, where the oilfields are "arguably one of the most strategic centres of oil supply for the United States in the post 9/11 world of energy security," the Ogoni from the delta region where Nigeria's vast oil reserves lie have developed an insurgent movement. The Ogoni have not only been excluded from the wealth generated through Nigerian oil production on their land but have also incurred the costs of oil operations. Michael Watts explains how the multi-billion-dollar oil industry of Nigeria has not produced an increase in per capita annual income; in fact, income has fallen from $250 per capita in 1965 to $212 in 2004, with the number of Nigerians living on less than one dollar a day increasing from 36 percent in 1970 (19 million people) to more than 70 percent in 2000 (90 million people). And yet Nigeria is rich with oil reserves, and government oil revenues swelled from 66 million naira in 1970 to more than 10 billion naira in 1980. And yet only one percent of Nigerians enjoy 85 percent of these oil revenues, with approximately $100 billion of $400 billion in oil revenues consistently going "missing" since 1970. Meanwhile, oil companies operating in Nigeria have not only used military forces against "insurgents" but have also directly funded militant groups as a security strategy. Moreover, says Watts, "their corrupt practices of distributing rents to local community elites" has on the whole "contributed to an environment in which military activity [is] in effect encouraged and facilitated."[30]

At this point, I return to the epigraphs at the beginning of the chapter. Given how hostile George W. Bush was to multilateralism during his two terms in office, so much so that he withdrew from the Kyoto Protocol and was criticized by Senator John Kerry for being isolationist, it seems surprising that he was the driving force behind the President's Emergency Plan for AIDS Relief, which provided $18.8 billion in funding to combat HIV/AIDS, treating 19 million people, mainly Africans.[31] That is, it is surprising until one starts connecting the dots of objective violence. Here one quickly comes to realize that the money Bush provided to help relieve HIV/AIDS in Africa might have been provided because African oil is of central importance to the United States and is predicted to become even more important than Middle Eastern oil. Against this backdrop, Bush's funding of HIV/AIDS programs in Africa seems like just another form of geopolitical oil strategizing—an act of economic opportunism, pure and simple.

All in all, what do the blue waters of the Gulf of Mexico drunk on 205.8 million gallons of crude oil, the United States as the top oil-consuming country in the world as of 2009,[32] climate change, the approximately 2 million people who died in the second Sudanese civil war (1983–2005), the thousands of Ogoni protesters murdered by Nigerian military forces in 1995, and U.S. "humanitarian" HIV/AIDS funding for Africans hold in common? Oil capitalism.

In light of the systemic violence throughout the history of oil production, how can responsibility be taken for the truth that ecological oil disasters present? This truth paradoxically cannot be accessed through visual perception alone; it requires a choice be made to refuse to be sidetracked by subjective forms of violence that, as Žižek explains, are perpetrated by "social agents, evil individuals, disciplined repressive apparatuses [and] fanatical crowds."[33] The spectacle of ecological disaster fetishizes "nature," and in so doing it invokes a form of obedience to the subjective violence documented by the image. People direct their energies of rage toward an Other, all the while ignoring the real source of the violence: social organizations that perpetuate the consumption of large amounts of fossil fuel energy.

Although the spectacle of violence after the *Deepwater Horizon* disaster appears to have redistributed the investment of energies organizing the U.S. social field, producing a crack in the everyday life of the average oil-consuming person, by March 2011 dissensus had once more flipped in favor of energy supplies over and above environmental concerns. One Gallup Poll noted that the "significant uptick in Americans' choosing the environment over production" just after the *Deepwater Horizon* spill had "proved a short-lived reaction to the event," going on to summarize the situation thus: "And a record-high 41% now think the U.S. should emphasize production of fossil fuels as the preferred solution to the nation's energy problems, although a 48% plurality continues to favor conservation. The same poll showed increased public support for offshore oil drilling and oil exploration in Alaska."[34]

Gallup reported in March 2011 that in a list of fourteen issues, the issue that concerned Americans most was the economy (71 percent). Environmental concern ranked thirteenth, with 34 percent reporting they worried about the environment a "great deal."[35]

So where did all the disagreement go? It turned into consensus when the images of cleanup set in, when the U.S. economic recovery slowed, and, more pertinent, when a criminal (BP) had been found. God forbid the criminal should be us! Denial of this sort is the pinnacle of objective violence. It emerged when the incomprehensible images were inscribed with meaning, enabling the general public to divert its energies of outrage toward the cleanup effort and BP. The effect of this diversion is that public dissensus was disciplined.

The well was capped because people put technology, science, and will power to work to stabilize the oil gushing from below the depths of the sea and earth. It was not out of altruism that the well was successfully capped and chemical dispersants were poured into the sea to break up the oil. It was pure self-interest, triggered by humanity's instinct for self-preservation, that resulted in this solution, and it was ultimately an equally aggressive move when human ingenuity abated the forces of material life with the use of technology. The human inclination to utilize the energy of material life in tandem with the power of technology and science turned material life against itself and ultimately changed the course of history (the oil stopped flowing). The public, politicians, BP, and Transocean all let out a sigh of relief. Unlike when the carbon-offset company Planktos had announced that it intended to disperse iron dust into the ocean to absorb carbon, in this case nobody was screaming, "How dare they mess with nature!" And why should anyone make such a claim? The point is that humanity solved the puzzle because in large part it had given up on the idea of "nature": it had in effect successfully messed with nature.

In this instance, messing with nature was a form of realism. It entailed realistically embracing the antagonistic drive at the heart of the human condition so that we came to accept failure. That said, this realism was missing the criticality that comes from self-reflexive thinking, for as we messed with nature, we failed to change what motivates the human instinct for self-preservation. The insatiable economic opportunism driving objective forms of violence throughout the politics of climate and environmental change remained unquestioned and intact. Indeed, this opportunism has been the primary focus of this book. I have concentrated on the different ways in which inequities operate throughout climate change and environmental politics, the discourses and practices of which by and large constitute acts of human aggression. Economic opportunism motivates the violence perpetrated upon other species, ecosystems, future generations, the poor, and the environments in which we live. The neoliberal character that the axiomatic of capital has taken and how it is used throughout climate change and environmental justice discourse and policy are merely the effects of this violence. The key now is to have the courage to mess with the "nature" of the human condition and extract from it the insatiable greed that produces injustice, inequity, and exploitation.

Here's a thought, albeit most likely offensive for some: What if people refused to clean up the oil-drenched birds, the muck floating onto the beaches, and so on and so forth? What if this refusal was made not out of a sense of helplessness, but in protest against the three different forms of capitalism that produced the ecological catastrophe in the first place and that have been described throughout this book: neoliberal capitalism, climate capitalism, and oil capitalism? As abhorrent as this proposition sounds, such a decision "not to act" might just render visible what the cleanup conceals: a realistic image of the wrath of capital and humanity's overall complicity with the violence such wrath generates. Now that might very well put the emotional charge of disaster imagery to work in the service of change.

AFTERWORD

Liberals feel unworthy of their possessions. Conservatives feel
they deserve everything they've stolen.

—Mort Sahl[1]

A quick snapshot of the twenty-first century so far: an economic meltdown; a frantic sell-off of public land to the energy business as President George W. Bush exited the White House; a prolonged, costly, and
unjustified war in Iraq; the Greek economy in ruins; an escalation of global
food prices; bee colonies in global extinction; 925 million hungry reported
in 2010; as of 2005, the world's five hundred richest individuals with a
combined income greater than that of the poorest 416 million people, the
richest 10 percent accounting for 54 percent of global income; a planet on
the verge of boiling point; melting ice caps; increases in extreme weather
conditions; and the list goes on and on and on.[2] Sounds like a ticking time
bomb, doesn't it? Well it is.

It is shameful to think that massive die-outs of future generations will
put to pale comparison the 6 million murdered during the Holocaust; the
millions killed in two world wars; the genocides in the former Yugoslavia,
Rwanda, and Darfur; the 1 million left homeless and the 316,000 killed by the
2010 earthquake in Haiti. The time has come to wake up to the warning signs.[3]

The real issue climate change poses is that we do not enjoy the luxury
of incremental change anymore. We are in the last decade where we can

do something about the situation. Paul Gilding, the former head of Greenpeace International and a core faculty member of Cambridge University's Programme for Sustainability, explains that "two degrees of warming is an inadequate goal and a plan for failure," adding that "returning to below one degree of warming . . . is the *solution to the problem*."[4] Once we move higher than 2°C of warming, which is what is projected to occur by 2050, positive feedback mechanisms will begin to kick in, and then we will be at the point of no return. We therefore need to start thinking very differently right now.

We do not see the crisis for what it is; we only see it as an isolated symptom that we need to make a few minor changes to deal with. This was the message that Venezuela's president Hugo Chávez delivered at the COP15 United Nations Climate Summit in Copenhagen on December 16, 2009, when he declared: "Let's talk about the cause. We should not avoid responsibilities, we should not avoid the depth of this problem. And I'll bring it up again, the cause of this disastrous panorama is the metabolic, destructive system of the capital and its model: capitalism."[5]

The structural conditions in which we operate are advanced capitalism. Given this fact, a few adjustments here and there to that system are not enough to solve the problems that climate change and environmental degradation pose.[6] Adaptability, modifications, and displacement, as I have consistently shown throughout this book, constitute the very essence of capitalism. Capitalism adapts without doing away with the threat. Under capitalism, one deals with threat not by challenging it, but by buying favors from it, as in voluntary carbon-offset schemes. In the process, one gives up on one's autonomy and reverts to being a child. Voluntarily offsetting a bit of carbon here and there, eating vegan, or recycling our waste, although well intended, are not solutions to the problem, but a symptom of the free market's ineffectiveness. By casting a scathing look at the neoliberal options on display, I have tried to show how all these options are ineffective. We are not buying indulgences because we have a choice; choices abound, and yet they all lead us down one path and through the golden gates of capitalist heaven.

For these reasons, I have underscored everyone's implication in this structure—myself included. If anything, the book has been an act of outrage—outrage at the deceit and the double bind that the "choices" under capitalism present, for there is no choice when everything is expendable. There is nothing substantial about the future when all you can do is survive by facing the absence of your own future and by sharing strength, stamina, and courage with the people around you. All the rest is false hope.

In many respects, writing this book has been an anxious exercise because I am fully aware that reducing the issues of environmental degradation and climate change to the domain of analysis can stave off the institution of useful solutions. But in my defense I would also like to propose that each and every one of us has certain skills that can contribute to making the solutions that we introduce in response to climate change and environmental degradation more effective and more realistic. In light of that view, I close with the following proposition, which I mean in the most optimistic sense possible: our politics must start from the point that after 2050 it may all be over.

NOTES

INTRODUCTION: BUSINESS AS USUAL

1. Samuel Beckett to Tom Driver, Summer 1961, in Lawrence Graver and Raymond Federman, eds., *Samuel Beckett: The Critical Heritage* (London: Routledge, 1997), 219.
2. In Hebrew, the word *golem* means "shapeless, lifeless matter."
3. The basic constituent of Marxist theory, as Antonio Negri so fittingly remarks, is that "neither the concept of capital nor its historical variants would exist in the absence of a proletariat which, whilst being exploited by capital, is always the living labour that produces it." See Antonio Negri, "Communism: Some Thoughts on the Concept and Practice," in Costas Douzinas and Slavoj Žižek, eds., *The Idea of Communism* (London: Verso, 2010), 156.
4. Fredric Jameson, *A Singular Modernity: Essay on the Ontology of the Present* (London: Verso, 2002).
5. International Organization for Migration, "Migration Climate Change and the Environment," n.d., at http://www.iom.int/jahia/Jahia/complex-nexus, accessed June 30, 2011.
6. The name "Anthropocene" is a geological classification and was coined by ecologist Eugene Stoermer. There is some debate when the period actually began, with some arguing that it started with industrialization, and others proposing that it commenced after World War II in 1945. For the sake of expediency, I refer to the more widely agreed upon definition that sets the date at the onset of industrialization. See Jan Zalasiewicz, Mark Williams, Will Steffen, and Paul Crutzen, "The New World of the Anthropocene," *Environmental Science and Technology* 44, no. 7 (2010): 2228–2231.
7. Johan Rockstöm, Will Steffen, Kenvin Noone, Åsa Persson, F. Stuart Chapin III, Eric F. Lambin, Timothy M. Lenton, et al., "A Safe Operating Space for Humanity," *Nature* 461 (September 24, 2009), 473.

8. Global Carbon Project, "Carbon Budget 2010," December 5, 2010, at http://
   www.globalcarbonproject.org/carbonbudget/10/hl-full.htm#Atmospheric
   Emissions, accessed February 26, 2012.
9. James Hansen's *Storms of My Grandchildren: The Truth About the Coming Climate
   Catastrophe and Our Last Chance to Save Humanity* (London: Bloomsbury, 2009) is
   essential reading for those who wish to understand the science of climate change.
10. For an excellent overview of the sociological aspects of climate change, see Joane
    Nagel, Thomas Dietz, and Jeffrey Broadbent, *Workshop on Sociological Perspectives
    on Climate Change, May 30–31, 2008* (Washington, D.C.: National Science Foun-
    dation, 2009), at http://ireswb.cc.ku.edu/~crgc/NSFWorkshop/Readings/NSF_
    WkspReport_09.pdf, accessed July 5, 2011.
11. Johan Rockström, Will Steffen, Kenvin Noone, Åsa Persson, F. Stuart Chapin III, Eric
    F. Lambin, Timothy M. Lenton, et al., "Planetary Boundaries: Exploring the Safe
    Operating Space for Humanity," *Ecology and Society* 14, no. 2, art. 32 (2009), 2, at
    http://www.stockholmresilience.org/download/18.8615c78125078c8d3380002197/
    ES-2009-3180.pdf, accessed July 1, 2011.
12. David Bollier, *Silent Theft: The Private Plunder of Our Common Wealth* (New York:
    Routledge, 2002); Michael Hardt and Antonio Negri, *Commonwealth* (Cambridge,
    Mass.: Belknap Press of Harvard University Press, 2009).
13. Merill Lynch and Capgemini, *2011 World Wealth Report* (New York: Merill Lynch,
    2011), 4, at http://www.ml.com/media/114235.pdf, accessed July 4, 2011.
14. Credit Suisse, press release, Zurich, October 8, 2010, at https://www.credit-suisse.
    com/news/en/media_release.jsp?ns=41610, accessed July 4, 2011.

1. CLIMATE CAPITALISM

1. GHGs trap heat in the earth's atmosphere. They include CO_2, methane, nitrous
   oxide, and three fluorinated gases (sulfur hexafluoride, perfluorocarbons, and
   hydrofluorocarbons). Although GHGs are produced and absorbed through natural cycles such as photosynthesis, it is now widely accepted that an imbalance has
   occurred as a result of human activities producing more GHG emissions than the
   earth's carbon sinks can absorb. See James Hansen, *Storms of My Grandchildren: The
   Truth About the Coming Climate Catastrophe and Our Last Chance to Save Humanity* (New York: Bloomsbury, 2009); Intergovernmental Panel on Climate Change
   (IPCC), *Climate Change 2007: Impacts, Adaptation, and Vulnerability. Contribution
   of Working Group II to the Fourth Assessment Report of the Intergovernmental Panel
   on Climate Change*, ed. Martin L. Parry, Osvaldo F. Canziani, Jean P. Palutikof, Paul
   J. van der Linden, and Clair E. Hanson (Cambridge: Cambridge University Press,
   2007); James Lovelock, *The Vanishing Face of Gaia: A Final Warning* (New York:
   Basic Books, 2009); and Bill McKibben, *The End of Nature* (New York: Random
   House, 1989).
2. IPCC, *Climate Change 2007*, 1000.

3. International Scientific Congress on Climate Change, "Rising Sea Levels Set to Have Major Impacts Around the World," press release, March 10, 2009, at http://climate-congress.ku.dk/newsroom/rising_sealevels, accessed January 6, 2010.

4. Nicholas Stern, *The Economics of Climate Change: The Stern Review* (Cambridge: Cambridge University Press, 2007).

5. Alliance of Small Island States (AOSIS), *Declaration on Climate Change 2009*, September 21, 2009, at http://www.sidsnet.org/aosis/documents/AOSIS%20 Summit%20Declaration%20Sept%2021%20FINAL.pdf, accessed March 27, 2010.

6. Thomas Heyd pointed out to me that at the March 2009 International Alliance of Research Universities Sciences conference there was agreement that the temperature should not be allowed to exceed even 2°C above preindustrial levels. So the solution that Rajendra Pachauri gives later in this chapter—slowing or reversing climate change through green technologies and green industry—seems to be conservative.

7. The World Bank devised the categories of low-, middle-, and high-income countries to classify countries on the basis of per capita gross national income. The groups are low income, $995 or less; lower-middle income, $996–$3,945; upper-middle income, $3,946–$12,195; and high income, $12,196 or more. These amounts are fixed during the World Bank's fiscal year. See World Bank, "How We Classify Countries," n.d., available at http://data.worldbank.org/about/country-classifications, accessed January 11, 2011.

8. Henry Shue, "After You: May Action by the Rich Be Contingent Upon Action by the Poor?" *Indiana Journal of Global Legal Studies* 1, no. 2 (1994): 343–366.

9. E. Wesley and F. Peterson, "The Ethics of Burden Sharing in the Global Greenhouse," *Journal of Agricultural and Environmental Ethics* 11 (1999), 191.

10. Peter Singer, *One World* (New Haven: Yale University Press, 2004).

11. Robin Attfield, *Environmental Ethics: An Overview for the Twenty-First Century* (Cambridge, Mass.: Polity, 2003); and George Monbiot, *Heat: How to Stop the Planet from Burning* (Cambridge, Mass.: South End Press, 2007).

12. Monbiot, *Heat*, 16.

13. Global Commons Institute, home page, available at http://www.gci.org.uk, accessed April 2, 2009.

14. Peter Newell and Matthew Paterson, *Climate Capitalism: Global Warming and the Transformation of the Global Economy* (Cambridge: Cambridge University Press, 2010). Although I share many of Newell and Paterson's political positions, I am less optimistic than they are that economic growth and the capitalist system can move us in a new direction.

15. Nick Dallas, *Green Business Basics: 24 Lessons for Meeting the Challenges of Global Warming* (New York: McGraw Hill, 2009); Rajendra Pachauri, speech at the opening ceremony for the UN Framework Convention for Climate Change Intergovernmental Panel on Climate Change, December 1, 2008, at http://www.ipcc.ch/press/index.htm#, accessed March 4, 2009; and Stern, *The Economics of Climate Change*.

16. William McDonough and Michael Braungart, *Cradle to Cradle: Remaking the Way We Make Things* (New York: Northpoint Press, 2002).

17. L. Hunter Lovins and Boyd Cohen, *Climate Capitalism: Capitalism in the Age of Climate Change* (New York: Hill and Wang, 2011), 3, 11, 300.

18. The phrase *green free market* comes from Professor John Shellnhuber, director of the Potsdam Institute for Climate Impact Research and member of the Climate Congress 2009 team. See International Scientific Congress on Climate Change, "Climate Change: Global Risks, Challenges, and Decisions Congress: Researchers Present Newest Update on Climate Change Science," press release, June 18, 2009, at http://climatecongress.ku.dk/newsroom/synthesis_report, accessed March 27, 2010.

19. Pachauri, IPCC speech, December 1, 2008.

20. Adrian Parr, *Hijacking Sustainability* (Cambridge, Mass.: MIT Press, 2009).

21. Karl Marx explained: "The transformation of a sum of money into means of production and labour-power is the first phase of the movement undergone by the quantum of value which is going to function as capital. It takes place in the market, within the sphere of circulation. The second phase of the movement, the process of production, is complete as soon as the means of production have been converted into commodities whose value exceeds that of their component parts, and therefore contains the capital originally advanced plus a surplus-value. These commodities must then be thrown back into the sphere of circulation. They must be sold, their value must be realized in money, this money must be transformed once again into capital, and so on, again and again. This cycle, in which the same phases are continually gone through in succession, forms the circulation of capital." Karl Marx, *Capital*, vol. 1, trans. Ben Fowkes (London: Penguin Books, 1990), 709.

22. Netherlands Environmental Assessment Agency, "China Contributing Two Thirds to Increase in CO2 Emissions," press release, June 13, 2008, at http://www.pbl.nl/en/news/pressreleases/2008/20080613ChinacontributingtwothirdstoincreaseinCO2emissions.html, accessed February 25, 2010. See also Netherlands Environmental Assessment Agency, "China Now No.1 in CO2 Emissions; USA in Second Position," press release, June 19, 2007, at http://www.pbl.nl/en/news/pressreleases/2007/20070619Chinanowno1inCO2emissionsUSAinsecondposition.html, accessed February 25, 2009.

23. Susan Houseman, "Outsourcing, Offshoring, and Productivity Measurement in U.S. Manufacturing," *International Labour Review* 146 (2007): 61–80.

24. See David Harvey, *A Brief History of Neoliberalism* (Oxford: Oxford University Press, 2005).

25. Tamsin Blanchard, *Green Is the New Black: How to Change the World with Style* (New York: William Morrow, 2008).

26. Kate Sheppard, "BP's $93 Million Ad Blitz," *Mother Jones*, September 22, 2010, at http://motherjones.com/mojo/2010/09/bps-ad-blitz, accessed June 1, 2011.

27. British Petroleum (BP), "Our Values," at http://www.bp.com/sectiongenericarticle.do?categoryId=9027967&contentId=7050884, accessed September 6, 2009.

28. Paul Hawken, *Blessed Unrest: How the Largest Movement Came Into Being and Why No-one Saw It Coming* (New York: Viking, 2007).

29. Chris Reidy, "Colgate Will Buy Tom's of Maine: $100m Deal May Help Boost Sales of Leader in Natural Products Niche," *Boston Globe*, March 22, 2006, at http://www.boston.com/business/articles/2006/03/22/colgate_will_buy_toms_of_maine, accessed May 1, 2006; Louise Story, "Can Burt's Bees Turn Clorox Green?" *New York Times*, January 6, 2008.

30. McDonough and Braungart, *Cradle to Cradle*.

2. GREEN ANGELS OR CARBON COWBOYS?

1. For more on a cost–benefit approach to climate change, see David Maddison, "A Cost–Benefit Analysis of Slowing Climate Change," *Energy Policy* 23, nos. 4–5 (1995): 337–346; and William Nordhaus, "The Cost of Slowing Climate Change: A Survey," *Energy Journal* 12, no.1 (1991): 37–65.

2. U.S. National Oceanic and Atmospheric Administration, Earth System Research Laboratory data, at ftp://ftp.cmdl.noaa.gov/ccg/co2/trends/co2_annmean_mlo.txt, accessed May 30, 2011.

3. Scientists have calculated the CO_2 equivalent of a GHG based on its global-warming potential over a one-hundred-year period. CO_2 has a global-warming potential of 1; methane's potential is 23 times that of CO_2; and nitrous oxide's potential is 296 times that of CO_2. Methane and nitrous oxide are therefore commonly described as "low-hanging fruit" because reducing emissions from these gases impacts climate change more than reducing CO_2 despite the fact that on a ton-for-ton basis CO_2 is causing the most warming. See U.S. Environmental Protection Agency, "Glossary of Climate Change Terms," n.d., at http://www.epa.gov/climatechange/glossary.html, accessed June 3, 2011.

4. The Kyoto Protocol became legally binding on February 16, 2005.

5. "The Conference of the Parties shall define the relevant principles, modalities, rules and guidelines, in particular for verification, reporting and accountability for emissions trading. The Parties included in Annex B may participate in emissions trading for the purposes of fulfilling their commitments under Article 3. Any such trading shall be supplemental to domestic actions for the purpose of meeting quantified emission limitation and reduction commitments under that Article." United Nations, *Kyoto Protocol to the United Nations Framework Convention on Climate Change* (1998), Art. 17, p. 15, at http://unfccc.int/resource/docs/convkp/kpeng.pdf, accessed May 30, 2011.

6. The JI, as defined in Article 6 of the protocol, "allows a country with an emission reduction or limitation commitment under the Kyoto Protocol (Annex B Party) to earn emission reduction units (ERUs) from an emission-reduction or emission removal project in another Annex B Party, each equivalent to one tonne of CO_2, which can be counted towards meeting its Kyoto target." United Nations Framework Convention on Climate Change, "Joint Implementation," n.d., at http://

unfccc.int/kyoto_protocol/mechanisms/joint_implementation/items/1674.php, accessed May 30, 2011.

7. The CDM is defined in Article 12 of the protocol. United Nations, *Kyoto Protocol to the United Nations Framework Convention on Climate Change*, 11–12.

8. In these countries, there are CDM-certified projects registered by the CDM Executive Board as of May 30, 2011. See United Nations Framework Convention on Climate Change, "CDM-Certified Projects," n.d., at http://cdm.unfccc.int/Projects/registered.html, accessed May 30, 2011.

9. Michael Wara, "Is the Global Carbon Market Working?" *Nature* 445 (February 8, 2007), 595.

10. United Nations Framework Convention on Climate Change, "Clean Development Mechanism," n.d., at http://unfccc.int/kyoto_protocol/mechanisms/clean_development_mechanism/items/2718.php, accessed May 30, 2011. In this first commitment period, the UN Framework Convention on Climate Change predicts it will register more than 1,650 projects.

11. The CCX was founded by Richard Sandor, an economics professor at Northwestern University, in conjunction with a grant from the Joyce Foundation.

12. Chicago Climate Exchange, "Overview," n.d., at http://www.chicagoclimatex.com/content.jsf?id=821, accessed June 1, 2011.

13. The CCX and its affiliates were purchased by IntercontinentalExchange in July 2010.

14. At the 2007 EU Summit, EU leaders agreed to cut emissions by 20 percent from 1990 levels by 2020. In addition, the union committed to receiving 20 percent of all its energy from renewables.

15. European Commission, "Emissions Trading System," *Climate Action*, updated November 15, 2010, at http://ec.europa.eu/clima/policies/ets/index_en.htm, accessed June 1, 2011.

16. Ibid.

17. For more on selling carbon offsets, see Ricardo Bayon, Amanda Hawn, and Katherine Hamilton, eds., *Voluntary Carbon Markets: An International Business Guide to What They Are and How They Work* (London: Earthscan, 2009), for which Al Gore wrote the foreword. Among such carbon-offset companies are Atmosfair, Climate Friendly, Climate Care, Native Energy, Carbon Counter, My Climate, eBlue, Greenswitch, Green Seat, SilvaTree, and KlimaFa. In addition to everyday individuals, the list of investors in carbon offsets includes the corporate sector, celebrities (Al Gore, Jon Bon Jovi, and members of Coldplay, the Rolling Stones, and the Dixie Chicks, to name a few), and even sovereign states, such as Vatican City.

18. As Michael Jenkins and Ricardo Bayon point out in their introduction to *Voluntary Carbon Markets*, the U.S. acid rain scheme, when translated into the context of carbon control, does not effectively address the issue of scale. They write: "Protecting one species, one piece of land, one watershed may be important, but it is no longer enough. The solutions today need to be systemic, they need to change the way we do business, the way we eat, drink, sleep and think." Michael Jenkins and Ricardo Bayon, "Introduction," in Bayon, Hawn, and Hamilton, eds., *Voluntary Carbon Markets*, xxi.

19. Wara, "Is the Global Carbon Market Working?" 595, 596.

20. Doug Struck, "Carbon Offsets: How a Vatican Forest Failed to Reduce Global Warming," *Christian Science Monitor*, April 20, 2010, at http://www.csmonitor.com/Environment/2010/0420/Carbon-offsets-How-a-Vatican-forest-failed-to-reduce-global-warming, accessed June 4, 2011.

21. Matt Richtel, "Recruiting Plankton to Fight Global Warming," *New York Times*, May 1, 2007, at http://www.nytimes.com/2007/05/01/business/01plankton.html, accessed June 4, 2011.

22. As reported in the *Washington Post*, "Other groups have looked on the company with less indulgence. The Surface Ocean Lower Atmosphere Study, an international research group, said last month that 'ocean fertilization will be ineffective and potentially deleterious, and should not be used as a strategy for offsetting CO_2 emissions.' The International Maritime Organization scientific group, the Friends of the Earth and the World Wildlife Fund have condemned it. And a group called the Sea Shepherd Conservation Society said its own ship would monitor the Planktos vessel and possibly 'intercept' it." Steven Mufson, "Iron to Plankton to Carbon Credits," *Washington Post*, July 20, 2007, at http://www.washingtonpost.com/wp-dyn/content/article/2007/07/19/AR2007071902553.html, accessed June 4, 2011.

23. Philip Fearnside, Daniel Lashof, and Philip Moura-Costa, "Accounting for Time in Mitigating Global Warming," *Mitigation and Adaptation Strategies for Global Change* 2 (1999): 285–302; Philip Fearnside, "Time Preference in Global Warming Calculations: A Proposal for a Unified Index," *Ecological Economics* 41, no. 1 (April 2002): 21–31.

24. Suzlon Energy is the world's third-largest wind turbine manufacturer.

25. Ben Arnoldy, "Carbon Offsets: Green Project Offends Indian Farmers Who Lose Land to Windmills," *Christian Science Monitor*, April 20, 2010, at http://www.csmonitor.com/Environment/2010/0420/Carbon-offsets-Green-project-offends-Indian-farmers-who-lose-land-to-windmills, accessed June 1, 2011.

26. Chris Lang and Timothy Byakola, *A Funny Place to Store Carbon: UWA–FACE Foundation's Tree Planting Project in Mount Elgon National Park, Uganda* (Montevideo, Uruguay: World Rainforest Movement, December 30, 2006), at http://chrislang.org/2006/12/30/a-funny-place-to-store-carbon-chapter-3, accessed June 8, 2011. All quotes and material in the subsequent discussion of the Benet and Mount Elgon situation come from this source.

27. I am referring to the Benet as indigenous to the Mount Elgon forest based on the ruling by Justice J. B. Katutsi in the High Court of Uganda on October 27, 2005. He stated that the Benet people "are historical and indigenous inhabitants of the said areas which were declared a Wildlife Protected Area or National Park." Quoted in ibid.

28. Katherine Hamilton, Milo Sjardin, Molly Peters-Stanley, and Thomas Marcello, *State of the Voluntary Carbon Markets 2010* (New York: Ecosystem Marketplace and Bloomberg New Energy Finance, June 14, 2010).

29. World Bank, "Outlook Is for Steady but Slower Growth in 2011 and 2012," in *Global Economic Prospects 2011*, a report (Washington, D.C.: World Bank, January 12, 2011), at http://go.worldbank.org/5AYIR3UW70, accessed June 1, 2011.

30. Karl Marx, *Capital*, vol. 1, trans. Ben Fowkes (London: Penguin Books, 1990), 138–139.
31. Ibid., 190, 212.
32. Ibid., 163.
33. Jenkins and Bayon, "Introduction," xxii.
34. Marx did not specifically use the term *false consciousness*; Engels did in a letter to Fran Mehrings. The concept was later developed by the Marxist scholar Georg Lukács in *History and Class Consciousness: Studies in Marxist Dialectics* (Cambridge, Mass.: MIT Press, 1971).
35. David Harvey, *The Limits to Capital*, 2nd ed. (London: Verso, 2006), 114, 116.
36. Ibid., 117, parentheses added.
37. Larry Lohmann, "Climate Change Politics After Montreal: Time for a Change," *Foreign Policy in Focus*, Issues: Energy, January 9, 2006, at http://www.fpif.org/articles/climate_politics_after_montreal_time_for_a_change, accessed June 2, 2011.
38. For more on the incongruities between a green lifestyle and continuing to live a life of high consumption, see Thomas L. Friedman, "Live Bad, Go Green," *New York Times*, July 8, 2007; Amy Richardson, "Carbon Credits—Paying to Pollute?" *3rd Degree* 2, no. 7 (October 10, 2006), at http://3degree.cci.ecu.edu.au/articles/view/781, accessed June 4, 2011; John Russell, "Are Emissions Offsets a Carbon Con?" *Ethical Corporation*, April 1, 2007, at http://www.greenbiz.com/news/reviews_third.cfm?NewsID=34804, accessed June 1, 2011.

3. POPULATION

1. From Population Reference Bureau, "2010 World Population Data Sheet," at http://www.prb.org/Publications/Datasheets/2010/2010wpds.aspx, accessed July 6, 2011.
2. I stipulate "*human* population growth" for the simple reason that at the same time that the number of people on earth is growing, we are also experiencing a mass extinction of species, which is in turn leading to a dramatic biodiversity loss. The International Union for Conservation of Nature estimates that the current species extinction rate is between one thousand and ten thousand times higher than the background rate. See International Union for the Conservation of Nature, "Species Extinction: The Facts" (Red List), February 26, 2009, at http://cmsdata.iucn.org/downloads/species_extinction_05_2007.pdf, accessed July 6, 2011. Recent reports include: Brian O'Neill, F. Landis MacKellar, and Wolfgang Lutz, *Population and Climate Change* (Cambridge: Cambridge University Press, 2001); Martha Campbell, John Cleland, Alex Eze, and Ndola Prata, "Return of the Population Growth Factor," *Science* 315, no. 5818 (March 16, 2007): 1501–1502.
3. United Nations, "World Population to Exceed 9 Billion by 2050," press release, March 11, 2009, at http://www.un.org/esa/population/publications/wpp2008/press-release.pdf, accessed March 20, 2010.
4. Paul R. Ehrlich, *The Population Bomb* (New York: Ballantine Books, 1968); James Lovelock, *The Revenge of Gaia: Earth's Climate Crisis and the Fate of Humanity* (New York: Basic Books, 2006), 141.

5. This special report is included in Intergovernmental Panel on Climate Change (IPCC), *Climate Change 2007: Impacts, Adaptation, and Vulnerability. Contribution of Working Group II to the Fourth Assessment Report of the Intergovernmental Panel on Climate Change*, ed. Martin L. Parry, Osvaldo F. Canziani, Jean P. Palutikof, Paul J. van der Linden, and Clair E. Hanson (Cambridge: Cambridge University Press, 2007).

6. Lovelock claims that the "root of our problem with the environment" comes from a "lack of constraint on the growth of population." Lovelock, *The Revenge of Gaia*, 141.

7. Ibid.

8. China's one-child policy was not consistently implemented. In rural areas, birth rates remained at 2.5 children per female, and in 2002 the government exempted rural areas from the law. The government also sometimes permitted a couple to have another child if their first child was female, in this way reinforcing the cultural bias against girls. The policy does not apply to Hong Kong, Macau, or foreign residents. Other exceptions to the one-child policy are if the first child is disabled, if both parents are from an ethnic minority, if both parents are themselves "single" children, if the couple divorces and the other spouse has not had a child of his or her own, or if the couple lived overseas and returns to live in China but had more than one child while they were residents overseas.

9. Some also claim that the imbalance between the number of Chinese men and the number of women "may have increased mental health problems and socially disruptive behavior among men." See Terese Hesketh, Li Lu, and Zhu Wei Xing, "The Effects of China's One-Child Family Policy After 25 Years," *New England Journal of Medicine* 353 (September 15, 2005): 1171–1176, at http://content.nejm.org/cgi/content/full/353/11/1171, accessed March 31, 2010.

10. I thank Kenneth Surin for bringing the situation of female infanticide in India to my attention. See "India's Unwanted Girls," *BBC News South Asia*, May 22, 2011, at http://www.bbc.co.uk/news/world-south-asia-13264301, accessed June 22, 2011.

11. David Warwick, *Bitter Pills: Population Policies and Their Implementation in Eight Developing Countries* (Cambridge: Cambridge University Press, 1982).

12. Betsy Hartmann, *Reproductive Rights and Wrongs: The Global Politics of Population Control*, rev. ed. (Boston: South End Press, 1995). See also Marika Vicziany, "Coercion in a Soft State: The Family Planning Program of India, Part One: The Myth of Voluntarism," *Pacific Affairs* 55, no. 3 (1982): 373–402.

13. Paul R. Ehrlich and Anne H. Ehrlich, "The Population Explosion: Why We Should Care and What We Should Do About It," *Environmental Law* (December 22, 1997): 1187–1208.

14. Paul A. Murtaugh and Michael G. Schlax, "Reproduction and the Carbon Legacies of Individuals," *Global Environmental Change* 19 (2009): 14–20.

15. Ibid., 14.

16. John Bongaarts, Brian C. O'Neill, and Stuart R. Gaffin, "Global Warming Policy: Population Left Out in the Cold," *Environment* 39, no. 9 (November 1997), 41.

17. Garrett Hardin, "The Tragedy of the Commons," *Science* 162, no. 3859 (1968): 1243–1248.

18. A well-known thought experiment that tests the premise at the core of Hardin's position is the prisoner's dilemma. Players are presented with the following situation. You and your friend are going to commit a crime, and prior to the act you agree that if you are ever caught, you will not betray the other person to the police. The two of you are caught by police and placed in separate holding cells. Neither of you can communicate with the other. The police then present you with the following deal: if you confess to the crime and provide evidence against your friend, who remains silent, you will go free, and your friend will receive the full sentence (ten years). If neither you nor your friend confesses to the crime, both of you will receive a two-year sentence (insufficient evidence for a full sentence). If, however, both you and your friend confess, then each of you will receive a five-year sentence. The question posed to participants of this game is: What would you do? Do you trust your friend to stay quiet and honor the pact the two of you made? Responding to the problem in this way would be the "cooperative" approach. Or do you provide evidence against your friend so that you don't serve any time and thus act out of self-interest? The dilemma arises when each person has to make a choice on the basis of not fully knowing what the other person will do. The lesson that the game teaches is that individuals usually act out of self-interest when it comes to distributing costs and benefits.

19. Hartmann, *Reproductive Rights and Wrongs*, 142–143.

20. Quoted in United Nations Population Fund, *The State of World Population 2009* (New York: United Nations, 2009), 51, at http://www.unfpa.org/swp/2009, accessed March 31, 2010.

21. In this sense, I share the conclusion arrived at by Betsy Hartmann, who argues that the larger families of the poor are a symptom of "unequal distribution of resources and power." Hartman, *Reproductive Rights and Wrongs*, 62.

22. For an excellent analysis of ecological entanglement, see Timothy Morton, *The Ecological Thought* (Cambridge, Mass.: Harvard University Press, 2010).

23. Leiwen Jiang and Karen Hardee, "How Do Recent Population Trends Matter in Climate Change?" *Population Action International* 1, no. 1 (April 30, 2009): 7, at http://populationaction.org/wp-content/uploads/2012/01/population_trends_climate_change_FINAL.pdf, accessed March 20, 2010. Also of interest is Brian C. O'Neill, Michael Dalton, Regina Fuchs, Leiwen Jiang, Shonali Pachauri, and Katarina Zigova, "Global Demographic Trends and Future Carbon Emissions," *Proceedings of the National Academy of Sciences* 107, no. 41 (October 12, 2010): 17521–17526.

24. Brian Massumi, *Parables for the Virtual: Movement, Affect, Sensation* (Durham, N.C.: Duke University Press, 2002), 69, emphasis in original.

25. Rosi Braidotti, "Feminist Epistemology After Postmodernism: Critiquing Science, Technology, and Globalisation," *Interdisciplinary Science Reviews* 32, no. 1 (2007), 67.

26. Nancy Fraser has developed a robust theory that argues in favor of pragmatically integrating three dimensions of justice. First, resources have to be equally distributed. Second, we have to tackle the ways in which cultural values work to marginalize women. Third, we need to strive toward achieving parity of participation. See

Nancy Fraser, *Justice Interruptus: Critical Reflections on the "Postsocialist" Condition* (New York: Routledge, 1997).

27. Derrida writes: "A written sign, in the usual sense of the word, is a mark which remains, which is not exhausted in the present of its inscription, and which can give rise to an iteration both in the absence of, and beyond the presence of the empirically determined subject who, in a given context, has emitted or produced it." Jacques Derrida, "Signature, Event, Context," in *Margins of Philosophy*, trans. Alan Bass (Chicago: University of Chicago Press, 1972), 317. John Searle most famously criticized Derrida's philosophy as nihilistic.

28. Michael Hardt and Antonio Negri, *Empire* (Cambridge, Mass.: Harvard University Press, 2000).

29. Michael Hardt and Antonio Negri, *Commonwealth* (Cambridge, Mass.: Belknap Press of Harvard University Press, 2009), 132.

30. Massumi, *Parables for the Virtual*, 70, emphasis in the original.

31. Hardt and Negri, *Commonwealth*, 131.

32. Giovanni Arrighi has criticized Hardt and Negri for having an overly idealistic view of the *potentia* of the multitude. See Giovanni Arrighi, "Lineages of Empire," *Historical Materialism* 10, no. 3 (2002): 3–16.

33. Marx writes: "But was not the caste regime also a particular division of labour? Was not the regime of the guilds another division of labour? And is not the division of labour under the system of manufacture, which in England begins in the middle of the seventeenth century and comes to an end in the last part of the eighteenth, totally different from the division of labour in large-scale modern industry?" Karl Marx to P. V. Annenkov, Brussels, December 28, 1846, in Karl Marx and Friedrich Engels, *Correspondence 1846–1895* (New York: International, 1935), 9.

34. Gilles Deleuze and Félix Guattari, *Anti-Oedipus: Capitalism and Schizophrenia*, trans. Robert Hurley, Mark Seem and Helen R. Lane (Minneapolis: University of Minnesota Press, 1996), 33.

4. TO BE OR NOT TO BE THIRSTY

1. Christopher L. Sabine, Richard A. Feely, Nicolas Gruber, Robert M. Key, Kitack Lee, John L. Bullister, Rik Wanninkhof, et al., "The Oceanic Sink for Anthropocentric CO_2," *Science* 305, no. 5682 (July 2004): 367–371.

2. National Aeronautics and Space Administration (NASA), "Evidence: Climate Change, How Do We Know?" *Climate Change: Vital Signs of the Planet* (a NASA online journal), n.d., at http://climate.nasa.gov/evidence, accessed July 2, 2011.

3. Species extinction is currently estimated to be occurring at rate of one hundred to one thousand times more than natural. Species extinction also erodes ecosystem resilience. See Johan Rockstöm, Will Steffen, Kenvin Noone, Åsa Persson, F. Stuart Chapin III, Eric F. Lambin, Timothy M. Lenton, et al., "A Safe Operating Space for Humanity," *Nature* 461 (September 24, 2009), 474. For more on this topic, see International Union for Conservation of Nature, "Wildlife in a Changing World: An

Analysis of the 2008 IUCN Red List of Threatened Species," n.d., at http://data.iucn. org/dbtw-wpd/edocs/RL-2009-001.pdf, accessed July 6, 2011.

4. Meena Palaniappan and Peter H. Gleick, "Peak Water," in Pacific Institute (Peter H. Gleick and others), *The World's Water 2008–2009: The Biennial Report on Freshwater Resources* (Washington, D.C.: Island Press, 2008), 4–6, at http://www.worldwater. org/data20082009/ch01.pdf, accessed September 1, 2010.

5. "Consumptive uses of water only refer to uses of water that make that water unavailable for immediate or short-term reuse within the same watershed. Such consumptive uses include water that has evaporated, transpired, been incorporated into products or crops, heavily contaminated, or consumed by humans or animals" (Ibid., 7).

6. United Nations Environment Programme (UNEP), *Global Environmental Outlook 4: Environment for Development* (Malta: UNEP, 2007), 129.

7. United Nations Water, "Statistics, Graphs, and Maps," n.d., at http://www.unwater .org/statistics.html, accessed September 1, 2010.

8. It is predicted that the peak in population will be followed by a slight decline that sees the global population level off at 8.97 billion by 2300. See United Nations Department of Economic and Social Affairs/Population Division, *World Population to 2300* (New York: United Nations Department of Economic and Social Affairs, 2004), at http://www.un.org/esa/population/publications/longrange2/WorldPop-2300final.pdf, accessed May 4, 2010.

9. World Resources Institute, "Water: Critical Shortages Ahead?" n.d., at http://www .wri.org/publication/content/8261, accessed June 15, 2010.

10. "Billions Daily Affected by Water Crisis," *Water*, n.d., at http://water.org/water-crisis/one-billion-affected, accessed December 7, 2010.

11. Ibid., accessed May 5, 2010.

12. Access to improved water not only differs from country to country but changes over time as well.

13. Let us not forget the craze around drinking bottled water. If, as reported in *Mother Jones*, the Carlyle hotel in Manhattan pours only Fiji bottled water in its dog bowls, I think it safe to say that bottled water is without doubt a "craze." World Economic Forum Water Initiative, "Managing Our Future Water Needs for Agriculture, Industry, Human Health, and the Environment," draft for discussion at the World Economic Forum Annual Meeting, January 2009, 11, at http://www3.weforum.org/ docs/WEF_ManagingFutureWater%20Needs_DiscussionDocument_2008.pdf, accessed January 15, 2011.

14. Water Footprint Network, home page, at http://www.waterfootprint.org/?page= files/home, accessed May 5, 2010.

15. Arjen Y. Hoekstra, *Water Neutral: Reducing and Offsetting the Impacts of Water Footprints*, Value of Water Research Report Series no. 28, UNESCO IH-E Institute for Water Education (Delft, Netherlands: Delft University of Technology, March 2008), 5.

16. Oscar Olivera and Tom Lewis, eds., *¡COCHABAMBA!: Water Wars in Bolivia* (Cambridge, Mass.: South End Press, 2004); Marcela Olivera, "The Cochabamba Water Wars: Marcela Olivera Reflects on the Tenth Anniversary of the Popular Uprising Against Bechtel and the Privatization of the City's Water Supply,"interview by Amy Goodman, *Democracy Now*, April 19, 2010, at http://www.democracynow.org/2010/4/19/the_cochabamba_water_wars_marcella_olivera, accessed June 5, 2010.

17. Nick Buxton, "Economic Strings: The Politics of Foreign Debt," in Jim Schultz and Melissa Crane Draper, eds., *Dignity and Defiance: Stories from Bolivia's Challenge to Globalization* (Berkeley: University of California Press, 2008), 174.

18. Jim Schultz and Melissa Crane Draper, "Conclusion," in Schultz and Draper, eds., *Dignity and Defiance*, 293. As the chief trade negotiator for the Morales administration in Bolivia, Pablo Solón explained, "'Free' trade and debt are two sides of the same coin. Debt was used to impose structural adjustment programs that sought to privatize and generate benefits for multinational companies. Free trade agreements are used to lock in the rules to benefit multinationals. In some ways it is easier to get out of debt by paying it off, as Argentina has done, than it is to extract yourself from commitments within free trade agreements." Quoted in Buxton, "Economic Strings," 175.

19. Kenneth Surin pointed out to me that the privatization of water resources can be traced back to the nineteenth century, when the English industrial centers of Birmingham, Liverpool, and Manchester pumped water from Welsh lakes to meet their industrial needs.

20. Center for Public Integrity and International Consortium of Investigative Journalists, *Promoting Privatization* (Washington, D.C.: Center for Public Integrity and International Consortium of Investigative Journalists, 2003), at http://projects.publicintegrity.org/water/report.aspx?aid=45, accessed May 10, 2010.

21. World Bank, *Water Resources Management: A World Bank Policy Paper* (Washington, D.C.: World Bank, 1993). Maude Barlow and Tony Clarke predict the growth of a water cartel that consists of transnational corporations, the World Bank, and governments. See Maude Barlow and Tony Clarke, *Blue Gold: The Fight to Stop the Corporate Theft of the World's Water* (New York: New Press, 2002).

22. Center for Public Integrity, *Promoting Privatization*.

23. Ibid.

24. John Perkins, *Confessions of an Economic Hit Man* (New York: Plume, 2006), 20–21.

25. Barlow and Clarke, *Blue Gold*.

26. Public Citizen, *Veolia Environment: A Corporate Profile*, Water for All Campaign (Washington, D.C.: Public Citizen, February 2005), at http://www.citizen.org/documents/Vivendi-USFilter.pdf, accessed November 1, 2010.

27. For a detailed discussion of this situation, see Erik Swyngedouw, "Dispossessing H2O: The Contested Terrain of Water Privatization," *Capitalism, Nature, Socialism* 16, no. 1 (March 2005), 89.

28. The notion of the commons can be traced back to the British idea of Crown Land and systems of management used for land held in common as outlined in the British Charter of the Forest (first issued in 1217). Although not mentioned in Hardt and Negri's book *Commonwealth*, it is important to point out that the idea of a "common wealth" was established for the protection of common pastures, as discussed in David Bollier, *Silent Theft: The Private Plunder of Our Common Wealth* (New York: Routledge, 2002). For more on this history, see Peter Linbaugh, *The Magna Carta Manifesto: Liberties and Commons for All* (Berkeley: University of California Press, 2009).

29. Michael Hardt and Antonio Negri, *Commonwealth* (Cambridge, Mass.: Belknap Press of Harvard University Press, 2010), xiv.

30. Elinor Ostrom, *Governing the Commons: The Evolution of Institutions for Collective Action* (Cambridge: Cambridge University Press, 1990).

31. For the use of game theory in analyzing the commons, see T. R. Lewis and J. Cowens, *Cooperation in the Commons: An Application of Repetitious Rivalry* (Vancouver: University of British Columbia, 1983).

32. Elinor Ostrom, "Public Entrepreneurship: A Case Study in Ground Water Basin Management," Ph.D. diss., University of California, Los Angeles, 1965.

33. William Blomquist and Elinor Ostrom, "Institutional Capacity and the Resolution of a Commons Dilemma," *Policy Studies Journal* 5, no. 2 (1985): 383–393.

34. Elinor Ostrom, James Walker, and Roy Gardner, *Rules, Games, and Common-Pool Resources* (Ann Arbor: University of Michigan Press, 1994).

35. Elinor Ostrom, James Walker, and Roy Gardner, "Covenants with and Without a Sword: Self-Governance Is Possible," *American Political Science Review* 86, no. 2 (June 1992), 404.

36. Ostrom has more recently been studying the role of trust in endogenous institutional arrangements, in particular those that assign property rights. See James C. Cox, Elinor Ostrom, James M. Walker, Antonio Jamie Castillo, Eric Coleman, Robert Holahan, Michael Schoon, and Brian Steed, "Trust in Private and Common Property Experiments," *Southern Economic Journal* 75 no. 4 (2009): 957–975.

37. "The nations with the largest net water loss are the USA (92 Gm3/yr), Australia (57 Gm3/yr), Argentina (47 Gm3/yr), Canada (43 Gm3/yr), Brazil (36 Gm3/yr) and Thailand (26 Gm3/yr). . . . The main products behind the national water loss from the USA are oil-bearing crops and cereal crops." A. K. Chapagain, A. Y. Hoekstra, and H. G. Savenije, "Water Saving Through International Trade of Agricultural Products," *Hydrology and Earth System Sciences* 10 (2006), 460, at http://www.waterfootprint.org/Reports/Chapagain_et_al_2006.pdf, accessed January 15, 2011.

38. Arjen Y. Hoekstra, "A Review of Research on Saving Water Through International Trade, National Water Dependencies, and Sustainability of Water Footprints," in *Virtual Water Trade: Documentation of an International Expert Workshop July 3–4, 2006* (Frankfurt am Main: Institute for Social-Ecological Research, 2006), 13.

39. See Simon Critchley, *Infinitely Demanding: Ethics of Commitment, Politics of Resistance* (London: Verso, 2007).

40. Hardt and Negri, *Commonwealth*, 137, viii, 288, emphasis in the original.
41. Karl Marx to Kugelman (first name unknown), London, July 11, 1868, in Karl Marx and Friedrich Engels, *Correspondence 1846–1895* (New York: International, 1935), 246, emphasis in original.
42. United Nations Water, "Water Use," n.d., at http://www.unwater.org/statistics_use.html, accessed May 5, 2010.
43. This idea of equal representation is the third tier of Nancy Fraser's theory of social justice (alongside recognition and redistribution). It is also at the heart of Jacques Rancière's definition of democracy as equality and of his argument that politics commences when the unrepresented disrupts the dominant political order. See Nancy Fraser, *Justice Interruptus: Critical Reflections on the "Postsocialist" Condition* (New York: Routledge, 1997); and Jacques Rancière, *Disagreement: Politics and Philosophy*, trans. Julie Rose (Minneapolis: University of Minnesota Press, 1999).
44. David Harvey's tripartite notion of a complex geography—spatiotemporality, environment, and places and regions—has been especially helpful for my analysis here because it invites us to address critically how vertical and horizontal lines of governance are produced and at what scale. See David Harvey, *Cosmopolitanism and the Geographies of Freedom* (New York: Columbia University Press, 2009), 85.
45. Refer to table 5.2 in Arjen Y. Hoekstra and Ashok K. Chapagain, *Globalization of Water: Sharing the Planet's Freshwater Resources* (Malden, Mass.: Blackwell, 2008), and "Access to Safe Drinking Water by Country," in Pacific Institute (Peter H. Gleick and others), *The World's Water 2008–2009: The Biennial Report on Freshwater Resources*, data table 3 (Washington, D.C.: Island Press, 2008), 214 and 217, at http://www.worldwater.org/data20082009/Table3.pdf, accessed May 5, 2010.
46. I address this process of signification in the chapter on slums in my book *Hijacking Sustainability* (Cambridge, Mass.: MIT Press, 2009).
47. These issues regarding desalination are identified as serious areas of concern in Heather Cooley, Peter H. Gleick, and Gary Wolff, *Desalination, with a Grain of Salt: A California Perspective* (Oakland, Calif.: Pacific Institute, June 2006).
48. As Marx noted, "It is only the dominion of past, accumulated, materialized labor over immediate living labor that stamps the accumulated labor with the character of capital." Karl Marx, *Wage-Labor and Capital*, trans. Friedrich Engels (New York: International, 1933), 30.
49. Cooley, Gleick, and Wolff, *Desalination*, 11–12.
50. Connor Boals, "Drinking from the Sea," *Circle of Blue Water*, June 29, 2009, at http://www.circleofblue.org/waternews/2009/world/drinking-from-the-sea-demand-for-desalination-plants-increases-worldwide, accessed June 30, 2010.
51. Quoted in ibid.
52. At the end of *Governing the Commons*, Ostrom briefly recognizes that the majority of her book has not "addressed the individual differences that exist among individuals involved in an institutional-choice situation." She admits: "Benefits and costs have to be discovered and weighed by individuals using human judgment in highly

uncertain and complex situations that are made even more complex to the extent that others behave strategically." Ostrom, *Governing the Commons*, 210.

53. Cathy Green and Sally Baden, *Gender Issues in Water and Sanitation Projects in Mali*, briefing commissioned by the Japanese International Cooperation Agency (Sussex, U.K.: IDS, Bridge, 1994).

54. Fraser, *Justice Interruptus*.

55. T. Van Ingen and C. Kawau, *Involvement of Women in Planning and Management in Tanga Region, Tanzania* (Gland, Switzerland: International Union for Conservation of Nature and World Conservation Union, 2003).

56. Theorists clearly situated in the camp that favors state regulation include Naomi Klein and David Harvey.

57. C. Rodriguez, *Water Management in the Bolivarian Republic of Venezuela* (Washington, D.C.: Embassy of the Bolivarian Republic of Venezuela in the United States, March 2010).

58. World Bank, "Retracting Glacier Impacts Economic Outlook in Tropical Andes," April 23, 2008, at http://go.worldbank.org/W5C3YWZFG0, accessed June 5, 2010.

59. The idea of moving beyond the egoism of limited partialities and natural rights to the invention of a generous society is at the core of Deleuze's discussion of David Hume's empiricism. See Gilles Deleuze, *Pure Immanence: Essays on a Life*, trans. Anne Boyman (New York: Zone Books, 2001), 46–47.

5. SOUNDING THE ALARM ON HUNGER

1. Billy Walters, interviewed by Lara Logan, *60 Minutes*, CBS, January 16, 2011.

2. United Nations Food and Agriculture Organization (FAO), *The State of Food Insecurity in the World: Addressing Food Insecurity in Protracted Crisis* (Rome: UN FAO, 2010), at http://www.fao.org/publications/sofi/en, accessed July 10, 2011.

3. Mark Lynas, *Six Degrees: Our Future on a Hotter Planet* (Washington, D.C.: National Geographic, 2008), 25–31.

4. World Health Organization (WHO), *Protecting Health from Climate Change: Connecting Science, Policy, and People* (Geneva: WHO, 2009), 8, at http://whqlibdoc.who.int/publications/2009/9789241598880_eng.pdf, accessed July 6, 2011.

5. Jarrod R. Welch, Jeffrey R. Vincent, Maximilian Auffhammer, Piedad F. Moya, Achim Dobermann, and David Dawe, "Rice Yields in Tropical/Subtropical Asia Exhibit Large but Opposing Sensitivities to Minimum and Maximum Temperatures," *Proceedings of the National Academy of Sciences* 107, no. 33 (August 17, 2010): 14562–14567.

6. Intergovernmental Panel on Climate Change (IPCC), *Climate 2007: The Physical Science Basis. Contribution of Working Group I to the Fourth Assessment Report of IPCC* (Cambridge: Cambridge University Press, 2007); Stephen P. Long, Elizabeth A. Ainsworth, Andrew D. B. Leakey, Josef Nösberger, and Donald R. Ort, "Food For Thought: Lower-Than-Expected Crop Yield Stimulation with Rising CO_2 Concentrations," *Science* 312 (June 30, 2006): 1918–1921.

7. Liliana Hisas, *The Food Gap: The Impacts of Climate Change on Food Production: A 2020 Perspective* (Alexandria, Va.: Fundación Ecológica Universal U.S., January 2011), iii, at http://www.feu-us.org/images/The_Food_Gap.pdf, accessed July 9, 2011.

8. Aiguo Dai, "Drought Under Global Warming: A Review," *Wiley Interdisciplinary Reviews: Climate Change* 2, no. 1 (January–February 2011), 50.

9. Ibid., 46.

10. Brian Fagan, *The Great Climate Warming: Climate Change and the Rise and Fall of Civilizations* (New York: Bloomsbury Press, 2008).

11. United Nations Food and Agriculture Organization (FAO), *Climate Change and Food Security: A Framework Document* (Rome: UN FAO, 2008), 9.

12. WHO, *Protecting Health from Climate Change*, 7.

13. Ibid., 14.

14. Quoted in UN FAO, *Climate Change and Food Security*, 3.

15. Ibid., 5.

16. Christoph Bals, Sven Harmeling, and Michael Windfuhr, *Climate Change, Food Security, and the Right to Adequate Food* (Stuttgart: Diakonisches Werk, 2008), 24.

17. WHO, *Protecting Health from Climate Change*, 2.

18. Ibid., 11.

19. UN FAO, *Climate Change and Food Security*, 42.

20. J. Dumanski, R. Peiretti, J. R. Benites, D. McGarry, and C. Pieri, "The Paradigm of Conservation Agriculture," *Proceedings of the World Association of Soil and Water Conservation*, Paper no. P1-7 (August 31, 2006), 59.

21. Adam Barclay, "Conserving the Future," *Rice Today* 5, no. 4 (October–December 2006), 22.

22. A. Ismail and B. Manneh, "Benin: Africa Component of STRASA Project Launches Second Phase," *STRASA News* 4, nos. 1–2 (June 2011): 5–6.

23. R. M. Baltazar and M. H. Dar, "Stress-Tolerant Rice Seeds Get a Boost with NGO Multiplication in West Bengal," *STRASA News* 4, nos. 1–2 (June 2011): 6–8.

24. See also Debal Deb, *Beyond Developmentality: Constructing Inclusive Freedom and Sustainability* (London: Earthscan, 2009).

25. Raj Patel, *Stuffed and Starved: Markets, Power, and the Hidden Battle for the World Food System* (Brooklyn: Melville House, 2007), 139.

26. A conversation between Debal Deb and me took place on Thursday, July 21, 2011, in which Dr. Deb recounted several instances of assault and harassment.

27. Vandana Shiva, *Biopiracy: The Plunder of Nature and Knowledge* (Cambridge, Mass.: South End Press, 1997), 7.

28. "In relation to health, a rights-based approach means integrating human rights norms and principles in the design, implementation, monitoring, and evaluation of health-related policies and programmes. These include human dignity, attention to the needs and rights of vulnerable groups, and an emphasis on ensuring that health systems are made accessible to all. The principle of equality and freedom from discrimination is central, including discrimination on the basis of sex and gender roles. Integrating human rights into development also means empowering

poor people, ensuring their participation in decision-making processes which concern them and incorporating accountability mechanisms which they can access." WHO, "Human Rights–Based Approach to Health," at http://www.who.int/trade/glossary/story054/en/index.html, under "Trade, Foreign Policy, Diplomacy, and Health," accessed July 8, 2011.

29. Hisas, *The Food Gap*.
30. United Nations Environment Programme, "World Food Supply: Food from Animal Feed," n.d., at http://www.grida.no/publications/rr/food-crisis/page/3565.aspx, accessed July 10, 2011.
31. U.S. Agency for International Development (USAID), "USAID Responds to Global Food Crisis," May 22, 2009, at http://www.usaid.gov/our_work/humanitarian_assistance/foodcrisis, accessed July 10, 2011.
32. Oxfam, "Bold Action Needed Now from G20 Agricultural Ministers to Tackle Causes of Food Price Volatility," June 21, 2011, at http://www.oxfam.org/en/grow/pressroom/pressrelease/2011-06-21/g20-agricultural-ministers-food-price-volatility, accessed July 21, 2011.
33. Ibid.
34. Lester R. Brown, *Biofuels Blunder: Massive Diversion of U.S. Grain to Fuel Cars in Raising World Food Prices, Risking Political Instability*, briefing before the U.S. Senate Committee on Environment and Public Works, 108th Cong., 1st sess., June 13, 2003, made available by the Earth Policy Institute at http://www.earth-policy.org/press_room/C68/senateepw07, accessed July 8, 2011.
35. Ibid.
36. Pimentel and Patzek summarize their findings thus:

    Findings in terms of energy outputs compared with the energy inputs were:
    • Ethanol production using corn grain required 29% more fossil energy than the ethanol fuel produced.
    • Ethanol production using switchgrass required 50% more fossil energy than the ethanol fuel produced.
    • Ethanol production using wood biomass required 57% more fossil energy than the ethanol fuel produced.
    • Biodiesel production using soybean required 27% more fossil energy than the biodiesel fuel produced ([n]ote, the energy yield from soy oil per hectare is far lower than the ethanol yield from corn).
    • Biodiesel production using sunflower required 118% more fossil energy than the biodiesel fuel produced.

    See David Pimentel and Tad W. Patzek, "Ethanol Production Using Corn, Switchgrass, and Wood; Biodiesel Production Using Soybean and Sunflower," *Natural Resources Research* 14, no. 1 (March 2005), 65.
37. Earth Policy Institute, "U.S. Corn Production and Use for Fuel Ethanol 1980–2009," n.d., at http://www.earth-policy.org/datacenter/xls/book_pb4_ch2_6.xls, accessed July 8, 2011.

38. Manuel Roig-Franzia, "A Culinary and Cultural Staple in Crisis," *Washington Post*, January 27, 2007, at http://www.washingtonpost.com/wp-dyn/content/article/2007/01/26/AR2007012601896.html, accessed July 7, 2011.

39. National Corn Growers Association, *Understanding the Impact of Higher Corn Prices on Consumer Food Prices* (Chesterfield, Mo.: National Corn Growers Association, March 26, 2007), at http://eerc.ra.utk.edu/etcfc/sefix/dos/FoodCornPrices.pdf, accessed July 8, 2011; and Roig-Franzia, "A Culinary and Cultural Staple in Crisis."

40. Steven Zahniser and William Coyle, *U.S.–Mexico Corn Trade During the NAFTA Era: New Twists to an Old Story*, FDS-04D-01 (Washington, D.C.: U.S. Department of Agriculture, May 2004), 2, at http://ip.cals.cornell.edu/courses/iard602/2007spring/mexico/mexico/USMEX_Corn_Trade.pdf, accessed July 8, 2011.

41. Comisión Nacional para el Conocimiento y Uso de la Biodiversidad (CONABIO), list of indigenous corn varieties, at http://www.conabio.gob.mx/2ep/images/f/f2/2EP_maiz_lenguas_ind%C3%ADgenas.pdf,accessed July 10, 2011.

42. "Corn Still a Better Bet Than Wheat Says Goldman," *Agrimoney*, July 15, 2010, at http://www.agrimoney.com/news/corn-still-a-better-bet-than-wheat-says-goldman-1985.html, accessed July 19, 2010.

43. Tom Levitt, "Goldman Sachs Makes $1 Billion Profit on Food Price Speculation," *The Ecologist* 40, no. 6 (July 19, 2010), at http://www.theecologist.org/News/news_round_up/542538/goldman_sachs_makes_1_billion_profit_on_food_price_speculation.html, accessed July 1, 2011.

44. Scott Irwin, *Is Speculation by Long-Only Index Funds Harmful to Commodity Markets?* testimony before the U.S. House of Representatives Committee on Agriculture, 108th Cong., 2nd sess., July 20, 2008, at http://www.farmdoc.illinois.edu/irwin/research/House%20Ag%20Testimony,%20July%202008.pdf, accessed July 1, 2011; Truman is quoted in this source.

45. Ibid.

46. Bob Dinneen, "Speculation: How Paper Bushels, Not Ethanol, Drive Corn," *American News*, April 1, 2011, at http://articles.aberdeennews.com/2011-04-01/farmforum/29373269_1_corn-crop-linn-group-corn-market, accessed July 11, 2011.

6. ANIMAL PHARM

1. Agricultural Statistics Board, National Agriculture Statistics Services, and U.S. Department of Agriculture, "Poultry Slaughter 2009 Summary," February 2010, at http://usda.mannlib.cornell.edu/usda/current/PoulSlauSu/PoulSlauSu-02-25-2010.pdf, accessed September 30, 2010; Agricultural Statistics Board, National Agriculture Statistics Services, and U.S. Department of Agriculture, "Livestock Slaughter 2009 Summary," April 2010, at http://usda.mannlib.cornell.edu/usda/current/LiveSlauSu/LiveSlauSu-04-29-2010.pdf, accessed September 30, 2010.

2. Donna J. Haraway, *Simians, Cyborgs, and Women: The Reinvention of Nature* (New York: Routledge, 1991), 149.

3. Henning Steinfeld, Pierre Gerber, Tom Wassenaar, Vincent Castel, Mauricio Rosales, and Cees de Haan, *Livestock's Long Shadow: Environmental Issues and Options* (Rome: United Nations Food and Agriculture Organization, 2006), xxi.

4. Michel Foucault, *"Society Must Be Defended": Lecture at the Collège de France, 1975–1976*, trans. David Macey (New York: Picador, 2003), 241.

5. Michel Foucault, *The History of Sexuality: An Introduction*, vol. 1, trans. Robert Hurley (New York: Vintage Books, 1990), 135–150; Michael Hardt and Antonio Negri, *Commonwealth* (Cambridge, Mass.: Belknap Press of Harvard University Press, 2009).

6. Donald D. Stull and Michael J. Broadway, *Slaughterhouse Blues: The Meat and Poultry Industry in North America* (Belmont, Calif.: Thomson/Wadsworth, 2003), 158.

7. Mercy for Animals, "Ohio Dairy Farm Brutality," April–May 2010, at http://www.mercyforanimals.org/ohdairy, accessed September 30, 2010.

8. Mercy for Animals, "Maine Egg Farm Investigation," 2008–2009, at http://www.mercyforanimals.org/maine-eggs, accessed September 30, 2009.

9. Nancy Fraser, "Feminism, Capitalism, and the Cunning of History," *New Left Review* 56 (March–April 2009), 98–99.

10. Ibid., 108.

11. Edward N. Wolff, "The Wealth Divide: The Growing Gap in the United States Between the Rich and the Rest," *Multinational Monitor* 24, no. 5 (May 2003), at http://multinationalmonitor.org/mm2003/03may/may03interviewswolff.html, accessed February 18, 2012.

12. Floyd Norris, "Off the Charts: In '08 Downturn, Some Managed to Eke Out Millions," *New York Times*, July 24, 2010.

13. The term *income* refers to earnings from work, rents, interest, dividends, and royalties.

14. Edward N. Wolff, *Recent Trends in Household Wealth in the United States: Rising Debt and the Middle-Class Squeeze—an Update to 2007*, Working Paper no. 589 (Annandale-on-Hudson, N.Y.: Levy Economics Institute, Bard College, March 2010), at http://www.levyinstitute.org/pubs/wp_589.pdf, accessed September 12, 2010.

15. This argument is one that Agamben makes. See Giorgio Agamben, *Homo Sacer: Sovereign Power and Bare Life*, trans. Daniel Heller-Roazen (Stanford: Stanford University Press, 1998), 125–127.

16. Peter Singer, *Animal Liberation* (New York: New York Review, 1975).

17. Tom Regan, *The Case for Animal Rights* (Berkeley: University of California Press, 1983). Regan's argument is also one that is used by antiabortionists (of which Regan is one) who on the same grounds claim that the fetus's right to life trumps the woman's right to decide what happens in and to her own body.

18. Carol J. Adams, *The Sexual Politics of Meat: A Feminist–Vegetarian Critical Theory* (New York: Continuum, 1990). Aviva Cantor has also studied the linguistic connections between the oppression of women and the oppression of animals. See Aviva

Cantor, "The Club, the Yoke, and the Leash: What We Can Learn from the Way a Culture Treats Animals," *Ms.* 12, no. 2 (August 1983): 27–29. Another fascinating ecofeminist study that has contributed to this field is Greta Gaard, "Vegetarian Ecofeminism: A Review Essay," *Frontiers* 23, no. 3 (2002): 117–146.

19. Adams, *The Sexual Politics of Meat*, 202.

20. Marti Kheel, *Nature Ethics: An Ecofeminist Perspective* (Lanham, Md.: Rowman and Littlefield, 2008), 3.

21. Bob Torres, *Making a Killing: The Political Economy of Animal Rights* (Oakland, Calif.: AK Press, 2007), 27.

22. Ibid., 58.

23. In an effort to respond to these inconsistencies, Marti Kheel eats a raw-food diet.

24. Lierre Keith, *The Vegetarian Myth: Food, Justice, and Sustainability* (Oakland, Calif.: PM Press, 2009).

25. Karl Marx, *Grundrisse*, trans. Martin Nicolaus (London: Penguin, 1993), 410, emphasis in original.

26. Michael Broadway, "Meatpacking and the Transformation of Rural Communities: A Comparison of Brooks, Alberta, and Garden City, Kansas," *Rural Sociology* 72, no. 4 (December 2007), 563.

27. U.S. Government Accountability Office (GAO), *Workplace Safety in the Meat and Poultry Industry, While Improving, Could Be Further Strengthened*, a report to the Ranking Minority Member, Committee on Health, Education, Labor, and Pensions, U.S. Senate (Washington, D.C.: U.S. GAO, January 2005). See also Jennifer Dillard, "A Slaughterhouse Nightmare: Psychological Harm Suffered by Slaughterhouse Employees and the Possibility of Redress Through Legal Reform," *Geography Journal on Poverty Law & Policy* 15, no. 2 (Summer 2008), 392.

28. Large slaughterhouses have grown, and from 1974 to 1997 the number of smaller plants decreased by nine hundred: "[From] 1974 to 1997 the number of packing houses employing more than a 1000 workers doubled from 24 to 48. During the same period, plants employing less than a 1000 workers dropped by over 900, and total industry employment fell by nearly 21,000 workers." Broadway, "Meatpacking," 562.

29. Hester J. Lipscomb, Robin Argue, Mary Anne McDonald, John M. Dement, Carol A. Epling, Tamara James, Steve Wing, and Dana Loomis, "Exploration of Work and Health Disparities Among Black Women Employed in Poultry Processing in the Rural South," *Environmental Health Perspectives* 113, no. 12 (December 2005), 1834, at http://ehp03.niehs.nih.gov/article/fetchArticle.action?articleURI=info%3Adoi%2F 10.1289%2Fehp.7912, accessed June 1, 2011.

30. U.S. GAO, *Workplace Safety in the Meat and Poultry Industry*, 3.

31. Sara A. Quandt, Joseph G. Grzywacz, Antonio Marin, Lourdes Carrillo, Michael L. Coates, Bless Burke, and Thomas A. Arcury, "Illnesses and Injuries Reported by Latino Poultry Workers in Western North Carolina," *American Journal of Industrial Medicine* 49 (2006), 349.

32. The culture of violence among slaughterhouse workers has led Jennifer Dillard to make a strong case in favor of treating slaughterhouse workers for post-traumatic disorder. See Dillard, "A Slaughterhouse Nightmare."

33. Quoted in ibid., 402–403.

34. In 2005, the U.S. GAO reported that 43 percent of the workers in the meat and poultry industry were younger than age thirty-five. In addition, 65 percent of the workforce were male. U.S. GAO, *Workplace Safety in the Meat and Poultry Industry*, 3.

35. Amy Fitzgerald, "Spill-Over from 'The Jungle' into the Larger Community: Slaughterhouses and Increased Crime Rates," 22, paper presented at the 2007 American Sociological Association annual meeting, August 11, 2007, New York City, at http://www.allacademic.com//meta/p_mla_apa_research_citation/1/8/3/0/1/pages183018/p183018-24.php, accessed September 27, 2010. On the connection between slaughterhouse workers and domestic violence, see Gail Eisnitz, *Slaughterhouse: The Shocking Story of Greed, Neglect, and Inhuman Treatment Inside the U.S. Meat Industry* (Amherst, N.Y.: Prometheus Books, 1997). In their study of the industrialization of meat- and poultry-production systems, Donald Stull and Michael Broadway found that when "farm size increases, so does rural poverty." Stull and Broadway, *Slaughterhouse Blues*, 149.

36. Jamie Fellner and Lance Compa, *Immigrant Workers in the United States Meat and Poultry Industry* (New York: Human Rights Watch, December 15, 2005), 9–10.

37. Ibid., 8.

38. Ibid., 10.

39. Quandt et al., "Illnesses and Injuries," 349.

40. Clifton B. Luttrell, *The High Cost of Farm Welfare* (Washington, D.C.: Cato Institute, 1989), 58.

41. B. Delworth Gardner and Carole Frank Nuckton, "Factors Affecting Agricultural Land Prices," *California Agriculture* (1979): 4–6.

42. Steve Huntley, "Winter of Despair Hits the Farm Belt," *U.S. News & World Report* 100 (January 20, 1986): 21–23; Bob McBride, "Broken Heartland: Farm Crisis in the Midwest," *The Nation* 242 (February 8, 1986): 132–133.

43. A finding by the Economic Research Service further confirms this thesis: "The amount of debt held by farm operators has increased substantially since 1990 . . . [becoming] concentrated in fewer farm businesses." See J. Michael Harris, James Johnson, John Dillard, Robert Williams, and Robert Dubman, *The Debt Finance Landscape for U.S. Farming and Farm Businesses*, a report from the Economic Research Service, AIS-87 (Washington, D.C.: U.S. Department of Agriculture, November 2009), 1; the quote in the text also comes from this source (p. 1).

44. Hardt and Negri, *Commonwealth*, 132–136.

45. From 1982 to 1997, the swine industry in North Carolina increased the number of hogs it produced fivefold, all the while reducing the number of farms from approximately eleven thousand to three thousand. See Stull and Broadway, *Slaughterhouse Blues*, 58.

46. Don P. Blaney, *The Changing Landscape of U.S. Milk Production* (Washington, D.C.: U.S. Department of Agriculture, June 2002), at http://www.ers.usda.gov/Publications /SB978, accessed September 1, 2010.

47. U.S. Department of Agriculture and National Agriculture Statistics Service, "Milk Production," September 17, 2010, at http://usda.mannlib.cornell.edu/usda/current/ MilkProd/MilkProd-09-17-2010.pdf, accessed September 25, 2010.

48. For a terrific ecofeminist analysis of recombinant bovine growth hormone (rBGH), see Greta Gaard, "Milking Mother Nature: An Ecofeminist Critique of rBGH," *The Ecologist* 24, no. 6 (November–December 1994): 202–203.

49. Dolly was euthanized on February 14, 2006, after suffering from lung cancer and arthritis.

50. Steve Stice Lab, University of Georgia, "What's Hot in the Stice Lab," n.d., at http:// www.biomed.uga.edu/stice, accessed October 1, 2010.

51. S. M. Willadsen, R. E. Janzen, R. J. McAlister, B. F. Shea, G. Hamilton, and D. McDermand, "The Viability of Late Morulae and Blastocysts Produced by Nuclear Transplantation in Cattle," *Theriogenology* 35, no. 1 (January 1991): 161–170.

52. U.S. Food and Drug Administration, "Animal Cloning," April 26, 2010, athttp://www .fda.gov/AnimalVeterinary/SafetyHealth/AnimalCloning/default.htm, accessed October 1, 2010.

53. "The procedure of somatic cloning is associated with important losses during pregnancy and in the perinatal period, reducing the overall efficacy to less than 5% in most cases. A mean of 30% of the cloned calves die before reaching 6 months of age with a wide range of pathologies, including, for the most common, respiratory failure, abnormal kidney development, liver steatosis. Heart and liver weight in relation to body weight are also increased." P. Chavatte-Palmer, D. Remy, N. Cordonnier, C. Richard, H. Issenman, P. Laigre, Y. Heyman, and J. P. Mialot, "Health Status of Cloned Cattle at Different Ages," *Cloning Stem Cells* 6, no. 2 (August 2004), 94. Other reports that raise red flags regarding the health of milk and meat from cloned animals are: National Academy of Sciences, Board of Agriculture and Natural Resources, *Animal Biotechnology: Science Based Concerns* (Washington, D.C.: National Academies Press, 2002); Merritt McKinney, "Flawed Genetic 'Marking' Seen in Cloned Animals," Reuters Health, May 29, 2001.

54. Although the Bayh–Dole Act is credited with stimulating university–industry collaborations, it did not result in the democratization of research findings. Instead, it led to the mass privatization of public-research outcomes. The declining state and increasing cost of health care in the United States is a case in point. In 2009, the Center for American Progress reported on the failing U.S. health-care system, noting that the cost of health care per person has more than doubled since 1994. It also noted that consumer inflation has averaged 2.6 percent per year; since 1994, per person health-care expenditure in the country has on average risen 5.5 percent per year. Research Foundation Technology Transfer Office, Colorado State University, "What Is Bayh–Dole and Why Is It Important to Technology Transfer?" October 1999, at http://www.csurf.org/enews/bayhdole_403.html, accessed October 2,

2010; and Center for American Progress and Ben Furnas, "American Health Care Since 1994: The Unacceptable Status Quo," January 8, 2009, at http://www.americanprogress.org/issues/2009/01/health_since_1994.html, accessed October 2, 2010. For a detailed analysis of the impact of the Bayh–Dole Act on university research in the United States and the commercialization of university research, see David C. Mowery, Richard R. Nelson, Bhaven N. Sampat, and Arvids A. Ziedonis, "The Effects of the Bayh–Dole Act on U.S. University Research and Technology Transfer: An Analysis of Data from Columbia University, the University of California, and Stanford University," paper presented at the Kennedy School of Government, Harvard University, September 10–12, 1998.

55. Deborah Blum, "The Brave New World of Steve Stice," *Georgia Magazine* 83, no. 3 (June 2004), at http://www.uga.edu/gm/300/FeatBrave.html, accessed October 1, 2010.

56. As reported by the Colorado State University Research Foundation, "The Bayh–Dole act is also vital to the university as a whole. University gross licensing revenues exceeded $200M in 1991 and by 1992 that number had risen to $250M. In FY 2000, U.S. and Canadian institution and universities Gross Licensing Income is reported in the AUTM survey at $1.26 Billion." Research Foundation Technology Transfer Office, "What Is Bayh–Dole?"

57. Denise Gellene, *Biotech Companies Trying to Milk Cloning for Profit* (Berkeley: Center for Genetics and Society, December 16, 2001), at http://www.geneticsandsociety.org/article.php?id=115, accessed October 1, 2010.

58. "ProLinia Announces Collaboration with Smithfield Foods," *PR Newswire*, June 19, 2000, at http://www.thefreelibrary.com/ProLinia+Announces+Collaboration+with+Smithfield+Foods-a062794189, accessed September 28, 2010.

59. ViaGen, "ViaGen Acquires Livestock Pioneer ProLinia," press release, June 30, 2003, at http://www.viagen.com/news/viagen-acquires-livestock-pioneer-prolinia, accessed June 1, 2011.

60. Diane Martindale, "Burgers on the Brain," *New Scientist* 177, no. 2380 (February 1, 2003): 26–29.

61. "According to national studies, lunches meet requirements for nutrients such as protein, vitamins, calcium, and iron, but do not meet the required 30 percent limit for calories from fat." See U.S. Government Accountability Office (GAO), *School Lunch Program: Efforts Needed to Improve Nutrition and Encourage Healthy Eating* (Washington, D.C.: U.S. GAO, May 2003).

62. *Supreme Beef Processors, Inc v. United States Department of Agriculture*, Defendant-Appellant no. 00-11008, U.S. Court of Appeals, 5th Cir., December 6, 2001, at http://caselaw.findlaw.com/us-5th-circuit/1429779.html, accessed September 2, 2010.

63. Helena Bottemiller, "Purdue Chicken Nuggets Recalled for Plastic," *Food Safety News*, July 20, 2010, at http://www.foodsafetynews.com/2010/07/91000-pounds-of-chicken-nuggets-recalled-for-plastic, accessed September 1, 2010

64. "Recall Expands to More Than Half a Billion Eggs," Associated Press, August 20, 2010, at http://www.msnbc.msn.com/id/38741401, accessed September 1, 2010.

65. Michael Pollan, *In Defense of Food: An Eater's Manifesto* (New York: Penguin Books, 2008), 122.

66. Raj Patel, *Stuffed and Starved: Markets, Power, and the Hidden Battle for the World Food System* (Brooklyn: Melville House, 2007), 1.

67. Raj Patel, *The Value of Nothing* (New York: Picador, 2009), 44–66.

68. Barbara Godoftas, "To Make a Tender Chicken," *Dollars & Sense* (July–August 2002): 14–30.

69. As noted by Heather Lipscomb and her colleagues, modern U.S. poultry-processing plants employ "large numbers of black and Hispanic women." Lipscomb et al., "Exploration of Work and Health Disparities," 1834.

70. These ideas are drawn heavily from the work of feminist philosopher of corporeality Elizabeth Grosz. See Elizabeth Grosz, *Volatile Bodies: Toward a Corporeal Feminism* (Bloomington: Indiana University Press, 1994).

71. Steinfeld et al., *Livestock's Long Shadow*, xxi.

72. Grosz, *Volatile Bodies*.

73. Hardt and Negri, *Commonwealth*, 133, emphasis in original.

7. MODERN FEELING AND THE GREEN CITY

1. Richard M. Daley's father, Richard J. Daley, was mayor of Chicago from 1955 to 1976.

2. Portland ranked first, San Francisco second, and Seattle third. SustainLane, "U.S. Sustainable City Rankings," n.d., at http://www.sustainlane.com/us-city-rankings/overall-rankings, accessed September 26, 2009.

3. Fredric Jameson, *Postmodernism, Or, The Cultural Logic of Late Capitalism* (Durham, N.C.: Duke University Press, 1991), 310.

4. David Harvey, *Spaces of Capital: Towards a Critical Geography* (Edinburgh: Edinburgh University Press, 2001), 333.

5. International Energy Agency, *Key World Energy Statistics 2010* (Paris: International Energy Agency, 2010), 30, at http://www.iea.org/textbase/nppdf/free/2010/key_stats_2010.pdf, accessed April 26, 2011.

6. Irving Mintzer, J. Amber Leonard, and Iván Dario Valencia, *Counting the Gigatonnes: Building Trust in Greenhouse Gas Inventories from the United States and China* (Washington, D.C.: World Wildlife Fund, June 2010, revised September 2010), vi.

7. Refer to figure 2.2 in United Nations Environment Programme (UNEP), *Buildings and Climate Change: Status, Challenges, and Opportunities* (Malta: UNEP, 2007), 5.

8. U.S. Green Building Council, "Buildings and Climate Change," n.d., at http://www.usgbc.org/DisplayPage.aspx?CMSPageID=2124, accessed April 1, 2011.

9. I have chosen the term *green cities* for the sake of expediency. Many other terms by and large refer to the same thing—*ecocity, sustainable city, environmentally friendly city*.

10. The main energy source for high- and middle-income areas is fossil fuels. In low-income areas such as rural India and China and in the African nations, it is usually

biomass (animal dung, wood, crop waste), kerosene, and paraffin. Many poor rural communities in low- and middle-income countries rely on wood for cooking purposes. This practice unfortunately contributes to the problem of desertification and deforestation (forest are important carbon sinks) as well as to health problems from the inhalation of smoke.

11. Per capita figures come from the United Nations Statistics Division, *Environmental Indicators: Greenhouse Gas Emissions 2007* (New York: United Nations Statistics Division, 2007), at http://unstats.un.org/unsd/environment/air_co2_emissions.htm, accessed April 26, 2011.

12. Pew Hispanic Center, *US Population Projections: 2000–2050* (Washington, D.C.: Pew Hispanic Center, February 11, 2008), at http://pewhispanic.org/reports/report. php?ReportID=85, accessed April 2, 2011.

13. The USGBC was founded by Mike Italiano, David Gottfried, and Rick Fedrizzi.

14. Michael Zaretsky, "LEED After Ten Years," in Adrian Parr and Michael Zaretsky, eds., *New Directions in Sustainable Design* (London: Routledge, 2010), 191.

15. Ibid.

16. The USGBC mission statement is quoted in ibid., 192.

17. Ibid., 193, 200.

18. Congress of New Urbanism, "The Charter of the New Urbanism," 1996, at http://www.cnu.org/charter, accessed April 20, 2011. See also Peter Calthorpe, *The Next American Metropolis: Ecology, Community, and the American Dream* (New York: Princeton Architectural Press, 1993); Andrés Duany and Elizabeth Plater-Zyberk, *Towns and Town-Making Principles* (New York: Rizzoli, 1991); and Peter Katz, *The New Urbanism: Toward an Architecture of Community* (New York: McGraw-Hill, 1994).

19. Michael Sorkin, "Can New Urbanism Learn from Modernism's Mistakes?" *Metropolis* 18, no. 1 (August–September 1998), 39.

20. Ibid.

21. Fred Davis, *Yearning for Yesterday: A Sociology of Nostalgia* (New York: Free Press, 1979); Kathleen Stewart, "Nostalgia—a Polemic," *Cultural Anthropology* 3, no. 3 (August 1988): 227–241.

22. Quoted in William Julius Wilson, *When Work Disappears: The World of the New Urban Poor* (New York: Vintage Books, 1997), 3–4.

23. Nicholas Lemann, *The Promised Land: The Great Black Migration and How It Changed America* (New York: Vintage, 1992).

24. U.S. Department of Housing and Urban Development, "About HOPE VI," 2009, at http://www.hud.gov/offices/pih/programs/ph/hope6/about, accessed April 26, 2011.

25. Loretta Lees, "Gentrification and Social Mixing: Towards an Inclusive Urban Renaissance?" *Urban Studies* 45, no. 12 (November 2008), 2449.

26. Pauline Lipman, "The Cultural Politics of Mixed-Income Schools and Housing: A Racialized Discourse of Displacement, Exclusion, and Control," *Anthropology & Education Quarterly* 40, no. 3 (2009), 218.

27. Ibid., 223.

28. Ibid., 222.

29. Charles Jencks, *The New Paradigm in Architecture: The Language of Post-Modernism* (New Haven: Yale University Press, 2002), 9. For a critique of Jencks, see Katherine Bristol, "The Pruitt-Igoe Myth," *Journal of Architectural Education* 44, no. 3 (May 1991): 163–171.

30. Jane Jacobs, *The Death and Life of Great American Cities* (New York: Random House, 1961).

31. Saskia Sassen, "A Global City," in Charles Madigan, ed., *Global Chicago* (Urbana: University of Illinois Press, 2004), 34.

32. Adele Simmons, "Introduction," in Madigan, ed., *Global Chicago*, 7–8, 14, emphasis added.

33. Douglas Farr, *Sustainable Urbanism: Urban Design with Nature* (Hoboken, N.J.: Wiley, 2008).

34. Philip Langdon, "A Booming Chicago Readies Itself for Rezoning," *New Urban Network* (March 2003), 7, at http://newurbannetwork.com/article/booming-chicago-readies-itself-rezoning, accessed April 22, 2011.

35. United Nations Human Settlement Programme (UN-HABITAT), *Hot Cities: Battleground for Climate Change* (Nairobi: UN-HABITAT, March 2011), at http://www.unhabitat.org/downloads/docs/GRHS2011/P1HotCities.pdf, accessed April 2, 2011.

36. *Chicago Climate Action Plan Report* (Chicago: City of Chicago, 2008), 6, at http://www.chicagoclimateaction.org/filebin/pdf/finalreport/CCAPREPORTFINAL.pdf, accessed September 26, 2009.

37. The statistics appear in Juliet Yonek and Romana Hasnain-Wynia, *A Profile of Health and Health Resources Within Chicago's 77 Community Areas* (Chicago: Northwestern University Feinberg School of Medicine, Center for Healthcare Equity/Institute for Healthcare Studies, 2011), 20, at http://chicagohealth77.org/uploads/Chicago-Health-Resources-Report-2011-0811.pdf, accessed February 20, 2012.

38. "Crime Statistics," *CityRating*, n.d., at http://www.cityrating.com/citycrime.asp?city=Chicago&state=IL, accessed April 6, 2011. On the deepening pattern of wage inequality throughout Chicago, see Marc Doussard, Jamie Peck, and Nik Theodore, "After Deindustrialization: Uneven Growth and Economic Inequality in 'Postindustrial' Chicago," *Economic Geography* 85, no. 2 (2009): 183–207.

39. Labor figures can be found in Doussard, Peck, and Theodore, "After Deindustrialization," 201, and are based on Chicago wage statistics from 1983 to 2004.

40. Gilles Deleuze and Félix Guattari, *Anti-Oedipus: Capitalism and Schizophrenia*, trans. Robert Hurley, Mark Seems, and Helen R. Lane (Minneapolis: University of Minnesota, 1977), 250.

41. David Harvey, *A Brief History of Neoliberalism* (Oxford: Oxford University Press, 2005).

42. Mari Gallagher Research and Consulting Group, *The Chicago Food Desert Progress Report* (Chicago: Mari Gallagher Research and Consulting Group, June 2009), 5, at http://www.marigallagher.com/site_media/dynamic/project_files/ChicagoFoodDesProg2009.pdf, accessed April 2, 2011.

43. Dick Simpson and Tom M. Kelly, "The New Chicago School of Urbanism and the New Daley Machine," *Urban Affairs Review* 44, no. 2 (November 2008), 232.

44. Lipman, "The Cultural Politics of Mixed-Income Schools and Housing," 218.

45. Karl Marx, *Capital*, vol. 1, trans. Ben Fowkes (London: Penguin Books, 1990), 137.

46. Chicago Convention and Tourist Bureau, "Choose Chicago," n.d., at http://www.choosechicago.com/media/statistics/visitor_impact/Pages/default.aspx, accessed April 29, 2011.

47. Chicago Office of Tourism, "2009 Statistical Information," n.d., 3, at http://www.explorechicago.org/etc/medialib/explore_chicago/tourism/pdfs_press_releases/chicago_office_of.Par.83640.File.dat/Statistics2006050708FINAL.pdf, accessed April 29, 2011.

48. David Harvey, *Limits to Capital*, 2nd ed. (London: Verso, 2006), 83.

8. SPILL, BABY, SPILL

1. "Transcript: Vice Presidential Debate," *New York Times*, October 2, 2008, at http://elections.nytimes.com/2008/president/debates/transcripts/vice-presidential-debate.html, accessed June 23, 2011.

2. Anthony Lake, Christine Todd Whitman, Princeton N. Lyman, and J. Stephen Morrison, *More Than Humanitarianism: A Strategic U.S. Approach Toward Africa* (New York: Council on Foreign Relations, 2006), xiii.

3. George W. Bush, *President's Emergency Plan for AIDS Relief (PEPFAR)* (Washington, D.C.: White House, 2003), at http://www.avert.org/pepfar.htm, accessed June 23, 2011.

4. In 2008, BP produced 271.4 million barrels of oil for the United States. See U.S. Energy Information Administration, *Oil: Crude and Petroleum Products Explained* (Washington, D.C.: U.S. Energy Information Administration, 2010), 6, at http://www.eia.gov/energyexplained/index.cfm?page=oil_home#tab2, accessed June 22, 2011.

5. See, among other sources on the connection between oil and violence, Amnesty International, *Oil in Sudan* (London: Amnesty International, 2001); Larry Everest, *Oil, Power, and Empire: Iraq and the U.S. Global Agenda* (New York: Common Courage Press, 2003); Michael T. Klare, *Blood and Oil: The Dangers and Consequences of America's Growing Dependency on Imported Oil* (New York: Metropolitan Books, 2004); Francisco Parra, *Oil Politics: A Modern History of Petroleum* (New York: I. B. Taurus, 2004).

6. John M. Broder, "Obama to Open Offshore Areas to Drilling for First Time," *New York Times*, March 31, 2010.

7. "Oil Spill Alters Views on Environmental Protection," Gallup Poll, May 27, 2010, at http://www.gallup.com/poll/137882/oil-spill-alters-views-environmental-protection.aspx, accessed June 22, 2011. It is interesting to compare data reported in a May 27, 2010, poll with data from a poll taken in April just prior to the spill. See Jeffrey M. Jones, "Americans Prioritize Energy Over the Environment for the First Time," Gallup Poll, April 6, 2010, at http://www.gallup.com/poll/127220/Americans-Prioritize-Energy-Environment-First-Time.aspx, accessed June 22, 2011.

8. Andrew J Hoffman and P. Devereaux Jennings, "The BP Oil Spill as a Cultural Anomaly? Institutional Context, Conflict, and Change," *Journal of Management Inquiry* 20, no. 2 (2011), 101.

9. That is, the violence of shock can be defused unless, as Slavoj Žižek so fittingly remarks in his use of Hegel, "there is something violent in the very symbolisation of a thing. . . . Language simplifies the designated thing, reducing it to a single feature. It dismembers the thing, destroying its organic unity. . . . It inserts the thing into a field of meaning which is ultimately external to it." Slavoj Žižek, *Violence* (New York: Picador, 2008), 61.

10. Hoffman and Jennings, "The BP Oil Spill," 109.

11. Bill McKibben, "Oil Spill Is an Opportunity for Americans," *U.S. News & World Report*, June 28, 2010, podcast at http://www.usnews.com/news/best-leaders/articles/2010/06/28/bill-mckibben-oil-spill-is-an-opportunity-for-americans, accessed June 20, 2011.

12. Sigmund Freud first used the term *disavowal* (*Verleugnung*) in 1914 in the Wolf Man case study.

13. William R. Freudenburg and Robert Gramling, *Blowout in the Gulf: The BP Oil Spill Disaster and the Future of Energy in America* (Cambridge, Mass.: MIT Press, 2011), 7.

14. Žižek's work on the condition of alterity posed by the violence of an event is especially helpful in my thinking here. He writes: "Though it may appear that there is a contradiction between the way discourse constitutes the very core of the subject's identity and the notion of this core as an unfathomable abyss beyond the 'wall of language,' there is a simple solution to this apparent paradox. The 'wall of language' which forever separates me from the abyss of another subject is simultaneously that which opens up and sustains the abyss—the very obstacle that separates me from the Beyond is what creates its mirage." Žižek, *Violence*, 73.

15. Bill McKibben, "Beyond Oil: Activism and Politics," *CounterCurrents*, August 27, 2010, at http://www.countercurrents.org/mckibben270810.htm, accessed June 22, 2011.

16. Naomi Klein, *The Shock Doctrine: The Rise of Disaster Capitalism* (New York: Picador, 2008).

17. In this part of the discussion, I am leaning on Rancière's concept of *le partage de sensible* (distribution or partition or sharing of the sensible). Such an interruption into everyday life completely reconfigures the sensible field people share in common with others around the world. By "sensible," Rancière means that which can be apprehended through perception and the senses. The rules, regimes, and hierarchies that constitute the social landscape distribute the sensible as much as aesthetics dislodges the sensible from these configurations, producing openings through which the previously excluded emerges into view. The image of ecological disaster, as Rancière noticed in the work of Flaubert, "asserts a molecular equality of affects that stands in opposition to the molar equality of subjects constructing a democratic political scene." Aesthetics, as sensory togetherness, is intrinsic to democratic politics because as the sensorium transmits the political energies of the social field, democracy takes on new meaning: it is "a form for constructing dissensus over 'the

given' of public life." Jacques Rancière, "The Janus-Face of Politicized Art: Jacques Rancière in Interview with Gabriel Rockhill," in *The Politics of Aesthetics*, trans. Gabriel Rockhill (London: Continuum, 2004), 56.

18. The incentive behind the Oil Pollution Act came from the Exxon Valdez oil spill in 1989, which was not long thereafter followed by oil spills off Rhode Island, the Delaware River, and the Houston Ship Channel within a three-month period. For more on the legal ramifications of the *Deepwater Horizon* blowout, see Robert Force, Martin Davies, and Joshua S. Force, "*Deepwater Horizon*: Removal Costs, Civil Damages, Crimes, Civil Penalties, and State Remedies in Oil Spill Cases," *Tulane Law Review* 85, no. 4 (March 2011): 889–982.

19. I should point out that I use the term *nature* reservedly, even ironically here, as a way to parenthesize the quasi-mystical view of nature as the backdrop for human activity and as an entity that exemplifies an image of harmony and purity in contradistinction to the artificial and alienated existence of human beings (especially Westerners).

20. See Rancière, *The Politics of Aesthetics*. Note that, for Marx, "men make their own history, but not of their own free will; not under circumstances they themselves have chosen but under the given and inherited circumstances with which they are directly confronted." Karl Marx, *The Eighteenth Brumaire of Louis Bonaparte, Surveys from Exile: Political Writings*, vol. 2, trans. and ed. David Fernbach (Harmondsworth, U.K.: Penguin, 1973), 146.

21. Timothy J. Crone and Maya Tolstoy, "Magnitude of the 2010 Gulf of Mexico Oil Leak," *Science* 330, no. 6004 (October 2010), 634.

22. Žižek, *Violence*, 9.

23. Hussein Mahdavy developed the meaning of the term *rentier state* forty years ago in "The Patterns and Problems of Economic Development in Rentier States: The Case of Iran," in M. A. Cook, ed., *Studies in Economic History of the Middle East*, 428–467 (London: Oxford University Press, 1970).

24. See Hazem Bablawi and Giacomo Luciani, eds., *The Rentier State* (New York: Croom Helm, 1987); John Clark, "Petro-Politics in Congo," *Journal of Democracy* 8, no. 3 (July 1997): 62–76; Douglas A. Yates, *The Rentier State in Africa: Oil Rent Dependency and Neocolonialism in the Republic of Gabon* (Trenton, N.J.: Africa World Press, 1996); and Kenneth Surin, "The Politics of the Southeast Asian Smog Crisis: A Classic Case of Rentier Capitalism at Work?" in Adrian Parr and Michael Zaretsky, eds., *New Directions in Sustainable Design*, 137–151 (London: Routledge, 2010).

25. For an excellent quantitative analysis of this phenomenon, see Michael Lewin Ross, "Does Oil Hinder Democracy?" *World Politics* 53, no. 3 (April 2001): 325–361.

26. United Nations High Commissioner for Refugees (UNHCR), U.S. Committee for Refugees and Immigrants, *U.S. Committee Mid Year Country Report—Sudan* (Geneva: UNHCR, October 2, 2001), at http://www.unhcr.org/refworld/country,,USCRI,,SDN,456d621e2,3c56c1161c,0.html, accessed June 23, 2011.

27. Ibid.

28. In 1980, the Numeiri government changed the North/South border, bringing the "oil provinces under central government jurisdiction, effectively disenfranchising

the South" Gaafar Numeiri was removed from power when popular opposition to his government mounted; the democratically elected government was eventually ousted, though, after a military coup in 1989 under the leadership of General Omar al-Bashir. Jason Switzer, *Oil and Violence in Sudan* (Winnipeg: International Institute for Sustainable Development and International Union for Conservation of Nature–World Conservation Union Commission on Environmental, Economic, and Social Policy, April 15, 2002), 6.

29. Ibid.

30. Michael Watts, *Imperial Oil: The Anatomy of a Nigerian Oil Insurgency*, Economies of Violence Working Papers, Working Paper no. 17 (Berkeley: Institute of International Studies, University of California, 2008), 3, 11, 15.

31. Chris McGreal, "George Bush: A Good Man in Africa," *Guardian UK*, February 15, 2008, at http://www.guardian.co.uk/world/2008/feb/15/georgebush.usa, accessed June 23, 2011; see also Bush, *PEPFAR*.

32. Maureen Hoch, "New Estimate Puts Gulf Oil Leak at 205 Million Gallons," *PBS Newshour*, August 2, 2010, at http://www.pbs.org/newshour/rundown/2010/08/new-estimate-puts-oil-leak-at-49-million-barrels.html, accessed June 17, 2011. Data through 2009 for 217 countries indicate that the United States ranked number one in the world for oil consumption, consuming 18,771,000 barrels a day. China was ranked second, consuming approximately 8,300,000 barrels a day. See U.S. Energy Information Administration, *Oil Consumption* (Washington, D.C.: U.S. Energy Information Administration, 2009), at http://www.eia.gov/countries/index.cfm?view=consumption, accessed June 23, 2011.

33. Žižck, *Violence*, 10.

34. Lydia Saad, "In U.S., Expanding Energy Output Still Trumps Green Concerns," Gallup Poll, March 16, 2011, at http://www.gallup.com/poll/146651/Expanding-Energy-Output-Trumps-Green-Concerns.aspx, accessed June 22, 2011.

35. See Lydia Saad, "Americans' Worries About Economy, Budget Top Other Issues," Gallup Poll, March 21, 2011, at http://www.gallup.com/poll/146708/Americans-Worries-Economy-Budget-Top-Issues.aspx, accessed June 22, 2011. The list of issues in order of ranking from highest to lowest were: environment, federal spending and the budget deficit, availability and affordability of health care, unemployment, Social Security system, size and power of the federal government, availability and affordability of energy, crime and violence, illegal immigration, hunger and homelessness, possibility of future terrorist attacks in the United States, drug use, quality of the environment, and race relations.

AFTERWORD: IN THE DANGER ZONE

1. This joke by stand-up comedian Mort Sahl is well known in the Jewish community.

2. United Nations Food and Agriculture Organization (UN FAO), *The State of Food Insecurity in the World: Addressing Food Insecurity in Protracted Crisis* (Rome: UN FAO, 2010), 8, at http://www.fao.org/docrep/013/i1683e/i1683e.pdf, accessed July 1,

2011; Kevin Watkins, *Summary Human Development Report 2005* (New York: United Nations Development Program, 2005), 18.

3. I thank my father, Mike Parr, whose views on climate changed helped shape my discussion here.

4. Paul Gilding, *The Great Disruption: Why the Climate Crisis Will Bring on the End of Shopping and the Birth of a New World* (New York: Bloomsbury Press, 2011), 128, emphasis in the original.

5. Hugo Chávez, President of the Bolivian Republic of Venezuela, speech at the COP15 United Nations Climate Summit, Copenhagen, December 16, 2009, at http://venezuela-us.org/live/wp-content/uploads/2009/12/16-DIC-09-DISCURSO-DEL-PRESIDENTE-CHAVEZ-EN COPENHAGUE-INGLÉS.pdf, accessed June 27, 2010.

6. Although the goal of this book has not been to provide a blueprint for change (several other books out there do this in a more informed way than my training allows), I can at least announce where I stand on the issue. I say stop the endless back and forth over whether to institute binding global agreements on emissions reductions or not and move ahead in disagreement. Ensure that GHG concentration in the atmosphere peaks by 2015 and remains at a global mean of less than 350 ppm CO_2e (as advised by Dr. James Hansen, who heads the NASA Goddard Institute for Space Studies; "CO_2e" refers to "all the main greenhouse gases converted into their equivalents in impact to CO_2, the key greenhouse gas of concern" [Gilding, *The Great Disruption*, 127]). Introduce a flat carbon tax the world over and apply serious penalties for those who do not comply. Keep fossil fuels in the ground. Revise immigration laws the world over in preparation for the mass exodus of environmental refugees. Governance on the basis of economic strength has to stop. Longer election cycles, especially in the United States, are needed to foster greater commitment toward long-term outcomes and more accountability. Social protection mechanisms have to be instituted across the board. Bring in annual restrictions on air miles traveled. Set up extensive public-transportation systems between and within all the world's metropolitan regions, and mandate a limit of one car per household. Introduce meat rationing. And last but not least, stop the privatization of the commons. On the topic of where we need to be with regard to GHG concentrations in the atmosphere, see James Hansen, Makiko Sato, Pushker Kharecha, David Beerling, Robert Berner, Valerie Masson-Delmotte, Mark Pagani, Maureen Raymo, Dana L. Royer, and James C. Zachos, "Target Atmospheric CO_2: Where Should Humanity Aim?" address to the NASA Goddard Institute for Space Studies, New York, October 15, 2008, at http://arxiv.org/pdf/0804.1126v3, accessed June 30, 2011.

BIBLIOGRAPHY

"Access to Safe Drinking Water by Country." In Pacific Institute (Peter H. Gleick and others), *The World's Water 2008–2009: The Biennial Report on Freshwater Resources*, data table 3. Washington, D.C.: Island Press, 2008. At http://www.worldwater.org/data20082009/Table3.pdf. Accessed May 5, 2010.

Adams, Carol J. *The Sexual Politics of Meat: A Feminist–Vegetarian Critical Theory*. New York: Continuum, 1990.

Agamben, Giorgio. *The Coming Community*. Trans. Michael Hardt. Minneapolis: University of Minnesota Press, 1993.

——. *Homo Sacer: Sovereign Power and Bare Life*. Trans. Daniel Heller-Roazen. Stanford: Stanford University Press, 1998.

——. *State of Exception*. Trans. Kevin Attell. Chicago: University of Chicago Press, 2005.

Agamben, Giorgio, Alain Badiou, Daniel Bensaid, Wendy Brown, Jean-Luc Nancy, Jacques Rancière, Kristin Ross, and Slavoj Žižek. *Democracy in What State?* Trans. William McCuaig. New York: Columbia University Press, 2011.

Agricultural Statistics Board, National Agriculture Statistics Services, and U.S. Department of Agriculture. "Livestock Slaughter 2009 Summary." April 2010. At http://usda.mannlib.cornell.edu/usda/current/LiveSlauSu/LiveSlauSu-04-29-2010.pdf. Accessed September 30, 2010.

——. "Poultry Slaughter 2009 Summary." February 2010. At http://usda.mannlib.cornell.edu/usda/current/PoulSlauSu/PoulSlauSu-02-25-2010.pdf. Accessed September 30, 2010.

Ali, Tariq, ed. *The Idea of Communism*. London: Seagull, 2009.

Alliance of Small Island States. *Declaration on Climate Change 2009*. September 21, 2009. At http://www.sidsnet.org/aosis/documents/AOSIS%20Summit%20Declaration%20Sept%2021%20FINAL.pdf. Accessed March 27, 2010.

Amnesty International. *Oil in Sudan*. London: Amnesty International, 2001.

Arendt, Hannah. *On Violence*. San Diego: Harcourt Brace, 1970.

Arnoldy, Ben. "Carbon Offsets: Green Project Offends Indian Farmers Who Lose Land to Windmills." *Christian Science Monitor,* April 20, 2010. At http://www.csmonitor .com/Environment/2010/0420/Carbon-offsets-Green-project-offends-Indian-farmers-who-lose-land-to-windmills. Accessed June 1, 2011.

Arputham, Jockin. "Recent Development Plans for Dharavi and for the Airport Slums in Mumbai." *Environment and Urbanization* 22, no. 2 (October 2010): 501–504.

Arrighi, Giovanni. "Lineages of Empire." *Historical Materialism* 10, no. 3 (2002): 3–16.

Attfield, Robin. *Environmental Ethics: An Overview for the Twenty-First Century.* Cambridge, Mass.: Polity, 2003.

Bablawi, Hazem and Giacomo Luciani, eds. *The Rentier State.* New York: Croom Helm, 1987.

Bals, Christoph, Sven Harmeling, and Michael Windfuhr. *Climate Change, Food Security, and the Right to Adequate Food.* Stuttgart: Diakonisches Werk, 2008.

Baltazar, R. M. and M. H. Dar. "Stress-Tolerant Rice Seeds Get a Boost with NGO Multiplication in West Bengal." *STRASA News* 4, nos. 1–2 (June 2011): 6–8.

Barclay, Adam. "Conserving the Future." *Rice Today* (October–December 2006): 18–23.

Barcott, Bruce. "Forlorn in the Bayou." *National Geographic* 218, no. 4 (October 2010): 62–75.

Barlow, Maude and Tony Clarke. *Blue Gold: The Fight to Stop the Corporate Theft of the World's Water.* New York: New Press, 2002.

Bayon, Ricardo, Amanda Hawn, and Katherine Hamilton, eds. *Voluntary Carbon Markets: An International Business Guide to What They Are and How They Work.* London: Earthscan, 2009.

Berube, Alan, Audrey Singer, and William Frey. *State of Metropolitan America.* Washington, D.C.: Metropolitan Policy Program, Brookings Institute, 2010. At http://www .brookings.edu/~/media/Files/Programs/Metro/state_of_metro_america/metro_ america_report.pdf. Accessed March 21, 2011.

Bied-Charreton, Marc. "Integrating the Combat Against Desertification and Land Degradation Into Negotiations on Climate Change: A Winning Strategy." November 2008. At http://www.unccd.int/science/docs/non_paper_desertif_Climate_eng.pdf . Accessed March 31, 2010.

"Billions Daily Affected by Water Crisis." *Water,* n.d. At http://water.org/water-crisis/ one-billion-affected. Accessed May 5 and December 7, 2010.

Blanchard, Tamsin. *Green Is the New Black: How to Change the World with Style.* New York: William Morrow, 2008.

Blaney, Don P. *The Changing Landscape of U.S. Milk Production.* Washington, D.C.: U.S. Department of Agriculture, June 2002. At http://www.ers.usda.gov/Publications/ SB978. Accessed September 1, 2010.

Blomquist, William and Elinor Ostrom. "Institutional Capacity and the Resolution of a Commons Dilemma." *Policy Studies Journal* 5, no. 2 (1985): 383–393.

Blum, Deborah. "The Brave New World of Steve Stice." *Georgia Magazine* 83, no. 3 (June 2004). At http://www.uga.edu/gm/300/FeatBrave.html. Accessed October 1, 2010.

Boals, Connor. "Drinking from the Sea." *Circle of Blue Water*, June 29, 2009. At http://www.circleofblue.org/waternews/2009/world/drinking-from-the-sea-demand-for-desalination-plants-increases-worldwide. Accessed June 30, 2010.

Bollier, David. *Silent Theft: The Private Plunder of Our Common Wealth*. New York: Routledge, 2002.

Bongaarts, John, Brian C. O'Neill, and Stuart R. Gaffin. "Global Warming Policy: Population Left Out in the Cold." *Environment* 39, no. 9 (November 1997): 40–41.

Bottemiller, Helena. "Purdue Chicken Nuggets Recalled for Plastic." *Food Safety News*, July 20, 2010. At http://www.foodsafetynews.com/2010/07/91000-pounds-of-chicken-nuggets-recalled-for-plastic. Accessed September 1, 2010

Bourne, Joel K., Jr. "The Gulf of Oil." *National Geographic* 218, no. 4 (October 2010): 28–61.

Braidotti, Rosi. "Feminist Epistemology After Postmodernism: Critiquing Science, Technology, and Globalisation." *Interdisciplinary Science Reviews* 32, no. 1 (2007): 65–74.

——. *Transpositions: On Nomadic Ethics*. Cambridge: Polity Press, 2006.

Bristol, Katherine. "The Pruitt-Igoe Myth." *Journal of Architectural Education* 44, no. 3 (May 1991): 163–171.

British Petroleum (BP). "Our Values." At http://www.bp.com/sectiongenericarticle.do?categoryId=9027967&contentId=7050884. Accessed September 6, 2009.

Broadway, Michael. "Meatpacking and the Transformation of Rural Communities: A Comparison of Brooks, Alberta, and Garden City, Kansas." *Rural Sociology* 72, no. 4 (December 2007): 560–582.

Broder, John M. "Obama to Open Offshore Areas to Drilling for First Time." *New York Times*, March 31, 2010.

Brown, Lester R. *Biofuels Blunder: Massive Diversion of U.S. Grain to Fuel Cars in Raising World Food Prices, Risking Political Instability*. Briefing before U.S. Senate Committee on Environment and Public Works, 108th Cong., 1st sess., June 13, 2003. Made available by the Earth Policy Institute at http://www.earth-policy.org/press_room/C68/senateepw07. Accessed July 8, 2011.

Brown, Wendy. *Regulating Aversion: Tolerance in the Age of Identity and Empire*. Princeton: Princeton University Press, 2008.

Buchanan, Ian. *Deleuzism: A Metacommentary*. Durham, N.C.: Duke University Press, 2000.

Bush, George W. *President's Emergency Plan for AIDS Relief (PEPFAR)*. Washington, D.C.: White House, 2003. At http://www.avert.org/pepfar.htm. Accessed June 23, 2011.

Buxton, Nick. "Economic Strings: The Politics of Foreign Debt." In Jim Schultz and Melissa Crane Draper, eds., *Dignity and Defiance: Stories from Bolivia's Challenge to Globalization*, 145–179. Berkeley: University of California Press, 2008.

Calthorpe, Peter. *The Next American Metropolis: Ecology, Community, and the American Dream*. New York: Princeton Architectural Press, 1993.

Campbell, Martha, John Cleland, Alex Eze, and Ndola Prata. "Return of the Population Growth Factor." *Science* 315, no. 5818 (March 16, 2007): 1501–1502.

Cantor, Aviva. "The Club, the Yoke, and the Leash: What We Can Learn from the Way a Culture Treats Animals." *Ms.* 12, no. 2 (August 1983): 27–29.

Castells, Manuel. "European Cities, the Informational Society, and the Global Economy." *New Left Review* 204 (March–April 1994): 18–32.

Center for American Progress and Ben Furnas. "American Health Care Since 1994: The Unacceptable Status Quo." January 8, 2009. At http://www.americanprogress.org/issues/2009/01/health_since_1994.html. Accessed October 2, 2010.

Center for Public Integrity and International Consortium of Investigative Journalists. *Promoting Privatization,* Washington, D.C.: Center for Public Integrity and International Consortium of Investigative Journalists, 2003 At http://projects.publicintegrity.org/water/report.aspx?aid=45. Accessed May 10, 2010.

Chapagain, A. K., A. Y. Hoekstra, and H. G. Savenije. "Water Saving Through International Trade of Agricultural Products." *Hydrology and Earth System Sciences* 10 (2006): 455–468. Also available at http://www.waterfootprint.org/Reports/Chapagain_et_al_2006.pdf. Accessed January 15, 2011.

Chavatte-Palmer P., D. Remy, N. Cordonnier, C. Richard, H. Issenman, P. Laigre, Y. Heyman, and J. P. Mialot. "Health Status of Cloned Cattle at Different Ages." *Cloning Stem Cells* 6, no. 2 (August, 2004): 94–100.

Chávez, Hugo, President of the Bolivian Republic of Venezuela. Speech at the COP15 United Nations Climate Summit, Copenhagen, December 16, 2009. At http://venezuela-us.org/live/wp-content/uploads/2009/12/16-DIC-09-DISCURSO-DEL-PRESIDENTE-CHAVEZ-EN-COPENHAGUE-INGLÉS.pdf. Accessed June 27, 2010.

Chicago Climate Action Plan Report. Chicago: City of Chicago, 2008. At http://www.chicagoclimateaction.org/filebin/pdf/finalreport/CCAPREPORTFINAL.pdf. Accessed September 26, 2009.

Chicago Climate Exchange. "Overview." n.d. At http://www.chicagoclimatex.com/content.jsf?id=821. Accessed June 1, 2011.

Chicago Convention and Tourist Bureau. "Choose Chicago." n.d. At http://www.choosechicago.com/media/statistics/visitor_impact/Pages/default.aspx. Accessed April 29, 2011.

Chicago Office of Tourism. "2009 Statistical Information." n.d. At http://www.explorechicago.org/etc/medialib/explore_chicago/tourism/pdfs_press_releases/chicago_office_of.Par.83640.File.dat/Statistics2006050708FINAL.pdf. Accessed April 29, 2011.

Clark, John. "Petro-Politics in Congo." *Journal of Democracy* 8, no. 3 (July 1997): 62–76.

Colomina, Beatriz. *Domesticity at War.* Cambridge, Mass.: MIT Press, 2007.

Colquhoun, Alan. *Modernity and the Classical Tradition: Architectural Essays 1980–1987.* Cambridge, Mass.: MIT Press, 1989.

Comisión Nacional Para el Conocimiento y Uso de la Biodiversidad (CONABIO). List of indigenous corn varieties. At http://www.conabio.gob.mx/2ep/images/f/f2/2EP_maiz_lenguas_ind%C3%ADgenas.pdf. Accessed July 10, 2011.

Congress of New Urbanism. "The Charter of the New Urbanism." 1996. At http://www
,cnu.org/charter. Accessed April 20, 2011.

Conley, Verena Andermatt. *Ecopolitics: The Environment in Poststructuralist Thought.*
London: Routledge, 1997.

Cook, M. A., ed. *Studies in Economic History of the Middle East.* London: Oxford University Press, 1970.

Cooke, Philip N., ed. *The Rise of the Rustbelt.* New York: St. Martin's Press, 1995.

Coole, Diana and Samantha Frost, eds. *New Materialisms: Ontology, Agency, and Politics.*
Durham, N.C.: Duke University Press, 2010.

Cooley, Heather, Peter H. Gleick, and Gary Wolff. *Desalination, with a Grain of Salt: A
California Perspective.* Oakland, Calif.: Pacific Institute, June 2006.

Cooper, Melinda. *Life as Surplus: Biotechnology & Capitalism in the Neoliberal Era.* Seattle: University of Washington Press, 2008.

"Corn Still a Better Bet Than Wheat Says Goldman." *Agrimoney,* July 15, 2010. At http://
www.agrimoney.com/news/corn-still-a-better-bet-than-wheat-says-goldman-1985
.html. Accessed July 19, 2010.

Cox, James C., Elinor Ostrom, James M. Walker, Antonio Jamie Castillo, Eric Coleman,
Robert Holahan, Michael Schoon, and Brian Steed. "Trust in Private and Common
Property Experiments." *Southern Economic Journal* 75, no. 4 (2009): 957–975.

Credit Suisse. Press release, Zurich, October 8, 2010. At https://www.credit-suisse.com/
news/en/media_release.jsp?ns=41610. Accessed July 4, 2011.

"Crime Statistics." *City Rating,* n.d. At http://www.cityrating.com/citycrime.asp?city=
Chicago&state=IL. Accessed April 6, 2011.

Critchley, Simon. *Infinitely Demanding: Ethics of Commitment, Politics of Resistance.*
London: Verso, 2007.

Croft, Catherine. "Movement and Myth: The Schröder House and Transformable Living." *Architectural Design* 70, no. 4 (2000): 10–15.

Crone, Timothy J. and Maya Tolstoy. "Magnitude of the 2010 Gulf of Mexico Oil Leak."
Science 330, no. 6004 (October 2010): 634.

Crumb, Michael J. "Egg Recall: US Chose Not to Require Vaccine for Salmonella Threat."
Huffington Post, August 24, 2010. At http://www.huffingtonpost.com/2010/08/25/
egg-recall-us-chose-not-t_n_693705.html. Accessed September 1, 2010.

Dai, Aiguo. "Drought Under Global Warming: A Review." *Wiley Interdisciplinary
Reviews: Climate Change* 2, no. 1 (January–February 2011): 45–65.

Dallas, Nick. *Green Business Basics: 24 Lessons for Meeting the Challenges of Global
Warming.* New York: McGraw Hill, 2009.

Davis, Fred. *Yearning for Yesterday: A Sociology of Nostalgia.* New York: Free Press, 1979.

Dear, Michael, ed. *From Chicago to L.A.: Making Sense of Urban Theory.* Thousand Oaks,
Calif.: Sage, 2002.

Deb, Debal. *Beyond Developmentality: Constructing Inclusive Freedom and Sustainability.* London: Earthscan, 2009.

Deleuze, Gilles. *Francis Bacon: The Logic of Sensation.* Trans. Daniel W. Smith. London:
Continuum, 2003.

——. *Pure Immanence: Essays on a Life*. Trans. Anne Boyman. New York: Zone Books, 2001.

Deleuze, Gilles, and Félix Guattari. *Anti-Oedipus: Capitalism and Schizophrenia*. Trans. Robert Hurley, Mark Seem, and Helen R. Lane. Minneapolis: University of Minnesota Press, 1996.

——. *A Thousand Plateaus: Capitalism and Schizophrenia*. Trans. Brian Massumi. London: Athlone Press, 1988.

——. *What Is Philosophy?* Trans. Graham Burchell and Hugh Tomlinson. London: Verso, 1994.

Derrida, Jacques. "Signature, Event, Context." In *Margins of Philosophy*, trans. Alan Bass, 307–330. Chicago: University of Chicago Press, 1972.

Diesendorf, Mark. *Climate Action: A Campaign Manual for Greenhouse Solutions*. Sydney: University of New South Wales Press, 2009.

Dillard, Jennifer. "A Slaughterhouse Nightmare: Psychological Harm Suffered by Slaughterhouse Employees and the Possibility of Redress Through Legal Reform." *Georgetown Journal on Poverty Law & Policy* 15, no. 2 (Summer 2008): 391–408.

Dinneen, Bob. "Speculation: How Paper Bushels, Not Ethanol, Drive Corn." *American News*, April 1, 2011. At http://articles.aberdeennews.com/2011-04-01/farmforum/29373269_1_corn-crop-linn-group-corn-market. Accessed July 11, 2011.

Doussard, Marc, Jamie Peck, and Nik Theodore. "After Deindustrialization: Uneven Growth and Economic Inequality in 'Postindustrial' Chicago." *Economic Geography* 85, no. 2 (2009): 183–207.

Douzinas, Costas and Slavoj Žižek, eds. *The Idea of Communism*. London: Verso, 2010.

Duany, Andrés and Elizabeth Plater-Zyberk. *Towns and Town-Making Principles*. New York: Rizzoli, 1991.

Dumanski, J., R. Peiretti, J. R. Benites, D. McGarry, and C. Pieri. "The Paradigm of Conservation Agriculture." *Proceedings of the World Association of Soil and Water Conservation*, Paper no. P1-7 (August 31, 2006): 58–64.

Dunham-Jones, Ellen and June Williamson. *Retrofitting Suburbia: Urban Design Solutions for Redesigning Suburbs*. Hoboken, N.J.: Wiley, 2009.

Earle, Sylvia. "The Blue Wilderness of My Childhood." *National Geographic* 218, no. 4 (October 2010): 76–77.

Earth Policy Institute. "US Corn Production and Use for Fuel Ethanol 1980–2009." n.d. At http://www.earth-policy.org/datacenter/xls/book_pb4_ch2_6.xls. Accessed July 8, 2011.

Ehrlich, Paul R. *The Population Bomb*. New York: Ballantine Books, 1968.

Ehrlich, Paul R. and Anne H. Ehrlich. "The Population Explosion: Why We Should Care and What We Should Do About It." *Environmental Law* (December 22, 1997): 1187–1208.

Eisnitz, Gail. *Slaughterhouse: The Shocking Story of Greed, Neglect, and Inhuman Treatment Inside the U.S. Meat Industry*. Amherst, N.Y.: Prometheus Books, 1997.

Engdahl, William. *A Century of War: Anglo-American Oil Politics and the New World Order*. London: Pluto Press, 2004.

European Commission. "Emissions Trading System." *Climate Action*, updated November 15, 2010, http://ec.europa.eu/clima/policies/ets/index_en.htm. Accessed June 5, 2011.

Everest, Larry. *Oil, Power, and Empire: Iraq and the U.S. Global Agenda*. New York: Common Courage Press, 2003.

Ewing, Reid, Tom Schmid, Richard Killingsworth, Amy Zlot, and Stephen Raudenbush. "Relationship Between Urban Sprawl and Physical Activity, Obesity, and Morbidity." *American Journal of Health Promotion* 18, no. 1 (September–October 2003): 47–57.

Fagan, Brian. *The Great Climate Warming: Climate Change and the Rise and Fall of Civilizations*. New York: Bloomsbury Press, 2008.

Farr, Douglas. *Sustainable Urbanism: Urban Design with Nature*. Hoboken, N.J.: Wiley, 2008.

Fearnside, Philip. "Time Preference in Global Warming Calculations: A Proposal for a Unified Index." *Ecological Economics* 41, no. 1 (April 2002): 21–31.

Fearnside, Philip, Daniel Lashof, and Philip Moura-Costa. "Accounting for Time in Mitigating Global Warming." *Mitigation and Adaptation Strategies for Global Change* 2 (1999): 285–302.

Fellner, Jamie and Lance Compa. *Immigrant Workers in the United States Meat and Poultry Industry*. New York: Human Rights Watch, December 15, 2005.

Fitzgerald, Amy. "Spill-Over from 'The Jungle' Into the Larger Community: Slaughterhouses and Increased Crime Rates." Paper presented at the 2007 American Sociological Association annual meeting, August 11, 2007, New York City. At http://www.allacademic.com//meta/p_mla_apa_research_citation/1/8/3/0/1/pages183018/p183018-24.php. Accessed September 27, 2010.

Flannery, Tim. *The Weather Makers: How Man Is Changing the Climate and What It Means for Life on Earth*. New York: Atlantic Monthly Press, 2005.

Force, Robert, Martin Davies, and Joshua S. Force. "*Deepwater Horizon*: Removal Costs, Civil Damages, Crimes, Civil Penalties, and State Remedies in Oil Spill Cases." *Tulane Law Review* 85, no. 4 (March 2011): 889–982.

"Foreclosure Activity Increases 4 Percent in July." *RealtyTrac*, August 12, 2010. At http://www.realtytrac.com/content/press-releases/foreclosure-activity-increases-4-percent-in-july-5946. Accessed March 3, 2011.

Foster, John Bellamy. *Marx's Ecology: Materialism and Nature*. New York: Monthly Review Press, 2000.

Foucault, Michel. *The History of Sexuality: An Introduction*. Vol. 1. Trans. Robert Hurley. New York: Vintage Books, 1990.

——. "*Society Must Be Defended*": Lectures at the Collège de France 1975–1976. Trans. David Macey. New York: Picador, 2003.

Fraser, Nancy. "Feminism, Capitalism, and the Cunning of History." *New Left Review* 56 (March–April 2009): 97–117.

——. *Justice Interruptus: Critical Reflections on the "Post Socialist" Condition*. New York: Routledge, 1997.

——. *Unruly Practices: Power, Discourse, and Gender in Contemporary Social Theory.* Minneapolis: University of Minnesota Press, 1989.

Freeman, Lance. "The Effects of Sprawl on Neighborhood Social Ties: An Explanatory Analysis." *Journal of the American Planning Association* 67, no. 1 (2001): 69–77.

Freudenburg, William R. and Robert Gramling. *Blowout in the Gulf: The BP Oil Spill Disaster and the Future of Energy in America.* Cambridge, Mass.: MIT Press, 2011.

Friedman, Thomas L. *Hot, Flat, and Crowded: Why We Need a Green Revolution—and How It Can Renew America.* New York: Farrar Straus and Giroux, 2008.

——. "Live Bad, Go Green." *New York Times*, July 8, 2007.

Gaard, Greta. "Milking Mother Nature: An Ecofeminist Critique of rBGH." *The Ecologist* 24, no. 6 (November–December 1994): 202–203.

——. "Vegetarian Ecofeminism: A Review Essay." *Frontiers* 23, no. 3 (2002): 117–146.

Galster, George, Royce Hanson, Hal Wolman, Stephen Coleman, and Jason Freihage. *Wrestling Sprawl to the Ground: Defining and Measuring an Elusive Concept.* Washington, D.C.: Fannie Mae Foundation, 2000. At http://content.knowledgeplex.org/kp2/programs/pdf/proc_fairgrowth_galster.pdf. Accessed March 3, 2011.

Gardner, B. Delworth and Carole Frank Nuckton. "Factors Affecting Agricultural Land Prices." *California Agriculture* (1979): 4–6.

Gellene, Denise. *Biotech Companies Trying to Milk Cloning for Profit.* Berkeley: Center for Genetics and Society, December 16, 2001. At http://www.geneticsandsociety.org/article.php?id=115. Accessed October 1, 2010.

Gilding, Paul. *The Great Disruption: Why the Climate Crisis Will Bring on the End of Shopping and the Birth of a New World.* New York: Bloomsbury Press, 2011.

Global Carbon Project. "Carbon Budget 2010." December 5, 2010. At http://www.global-carbonproject.org/carbonbudget/10/hl-full.htm#AtmosphericEmissions. Accessed February 26, 2012.

Global Commons Institute. Home page. At http://www.gci.org.uk. Accessed April 2, 2009.

Godoftas, Barbara. "To Make a Tender Chicken." *Dollars & Sense* (July–August 2002): 14–30.

Gökay, Bülent. *Politics of Oil: A Survey.* London: Routledge, 2006.

Graham, Stephen. *Cities Under Siege: The New Military Urbanism.* London: Verso, 2010.

Graver, Lawrence and Raymond Federman, eds. *Samuel Beckett: The Critical Heritage.* London: Routledge, 1997.

Green, Cathy and Sally Baden. *Gender Issues in Water and Sanitation Projects in Mali.* Briefing commissioned by the Japanese International Cooperation Agency. Sussex, U.K.: IDS, Bridge, 1994.

Grosz, Elizabeth. *Chaos, Territory, Art: Deleuze and the Framing of the Earth.* New York: Columbia University Press, 2008.

——. *Volatile Bodies: Toward a Corporeal Feminism.* Bloomington: Indiana University Press, 1994.

Guattari, Félix. *Chaosophy.* Ed. Sylvére Lotringer. New York: Semiotext(e), 1995.

Halle, David. *New York and Los Angeles: Politics, Society, and Culture, a Comparative View*. Chicago: University of Chicago Press, 2003.

Hamilton, Katherine, Milo Sjardin, Molly Peters-Stanley, and Thomas Marcello. *State of the Voluntary Carbon Markets 2010*. New York: Ecosystem Marketplace and Bloomberg New Energy Finance, June 14, 2010.

Hansen, James. *Storms of My Grandchildren: The Truth About the Coming Climate Catastrophe and Our Last Chance to Save Humanity*. New York: Bloomsbury, 2009.

Hansen, James, Makiko Sato, Pushker Kharecha, David Beerling, Robert Berner, Valerie Masson-Delmotte, Mark Pagani, Maureen Raymo, Dana L. Royer, and James C. Zachos. "Target Atmospheric CO2: Where Should Humanity Aim?" Address to the NASA Goddard Institute for Space Studies, New York, October 15, 2008. At http://arxiv.org/pdf/0804.1126v3. Accessed June 30, 2011.

Haraway, Donna J. *Simians, Cyborgs, and Women: The Reinvention of Nature*. New York: Routledge, 1991.

Hardin, Garrett. "The Tragedy of the Commons." *Science* 162, no. 3859 (1968): 1243–1248.

Hardt, Michael and Antonio Negri. *Commonwealth*. Cambridge, Mass.: Belknap Press of Harvard University Press, 2010.

——. *Empire*. Cambridge, Mass.: Harvard University Press, 2000.

——. *Multitude: War and Democracy in the Age of the Empire*. New York: Penguin, 2004.

Harris, J. Michael, James Johnson, John Dillard, Robert Williams, and Robert Dubman. *The Debt Finance Landscape for U.S. Farming and Farm Businesses*. AIS-87. Washington, D.C.: Economic Research Service, U.S. Department of Agriculture, November 2009.

Hartmann, Betsy. *Reproductive Rights and Wrongs: The Global Politics of Population Control*. Rev. ed. Boston: South End Press, 1995.

Harvey, David. *A Brief History of Neoliberalism*. Oxford: Oxford University Press, 2005.

——. *Cosmopolitanism and the Geographies of Freedom*. New York: Columbia University Press, 2009.

——. *The Limits to Capital*. 2nd ed. London: Verso, 2006.

——. *The New Imperialism*. Oxford: Oxford University Press, 2003.

——. "The Right to the City." *New Left Review* 53 (September–October 2008): 23–40.

——. *Spaces of Capital: Towards a Critical Geography*. Edinburgh: Edinburgh University Press, 2001.

——. *Spaces of Global Capitalism: Towards a Theory of Uneven Geographical Development*. London: Verso, 2006.

——. *The Urbanization of Capital: Studies in the History and Theory of Capitalist Urbanization*. Baltimore: Johns Hopkins University Press, 1985.

Hatkoff, Amy. *The Inner World of Farm Animals: Their Amazing Social, Emotional, and Intellectual Capacities*. New York: Stewart, Tabori & Chang, 2009.

Hawken, Paul. *Blessed Unrest: How the Largest Movement Came Into Being and Why No-one Saw It Coming*. New York: Viking, 2007.

Hesketh, Terese, Li Lu, and Zhu Wei Xing. "The Effects of China's One-Child Family Policy After 25 Years." *New England Journal of Medicine* 353 (September 15, 2005):

1171–1176. At http://content.nejm.org/cgi/content/full/353/11/1171. Accessed March 31, 2010.

Hisas, Liliana. *The Food Gap: The Impacts of Climate Change on Food Production: A 2020 Perspective.* Alexandria, Va.: Fundación Ecológica Universal U.S., January 2011. At http://www.feu-us.org/images/The_Food_Gap.pdf. Accessed July 9, 2011.

Hoch, Maureen. "New Estimate Puts Gulf Oil Leak at 205 Million Gallons." *PBS Newshour,* August 2, 2010. At http://www.pbs.org/newshour/rundown/2010/08/new-estimate-puts-oil-leak-at-49-million-barrels.html. Accessed June 17, 2011.

Hoekstra, Arjen Y. "A Review of Research on Saving Water Through International Trade, National Water Dependencies, and Sustainability of Water Footprints." In *Virtual Water Trade: Documentation of an International Expert Workshop, July 3–4, 2006,* 12–15. Frankfurt am Main: Institute for Social-Ecological Research, 2006.

——. *Water Neutral: Reducing and Offsetting the Impacts of Water Footprints.* Value of Water Research Report Series no. 28, UNESCO IH-E Institute for Water Education. Delft, Netherlands: Delft University of Technology, March 2008.

Hoekstra, Arjen Y. and Ashok K. Chapagain. *Globalization of Water: Sharing the Planet's Freshwater Resources.* Malden, Mass.: Blackwell, 2008.

Hoffman, Andrew J. and P. Devereaux Jennings. "The BP Oil Spill as a Cultural Anomaly? Institutional Context, Conflict, and Change." *Journal of Management Inquiry* 20, no. 2 (2011): 100–112.

Honneth, Axel. *Disrespect: The Normative Foundations of Critical Theory.* Cambridge: Polity, 2008.

Houseman, Susan. "Outsourcing, Offshoring, and Productivity Measurement in U.S. Manufacturing." *International Labour Review* 146 (2007): 61–80.

Huntley, Steve. "Winter of Despair Hits the Farm Belt." *U.S. News & World Report* 100 (January 20, 1986): 21–23.

"India's Unwanted Girls." *BBC News South Asia,* May 22, 2011. At http://www.bbc.co.uk/news/world-south-asia-13264301. Accessed June 22, 2011.

Intergovernmental Panel on Climate Change (IPCC). *Climate 2007: The Physical Science Basis. Contribution of Working Group I to the Fourth Assessment Report of IPCC.* Cambridge: Cambridge University Press, 2007.

——. *Climate Change 2007: Impacts, Adaptation, and Vulnerability. Contribution of Working Group II to the Fourth Assessment Report of the Intergovernmental Panel on Climate Change.* Ed. Martin L. Parry, Osvaldo F. Canziani, Jean P. Palutikof, Paul J. van der Linden, and Clair E. Hanson. Cambridge: Cambridge University Press, 2007.

International Energy Agency. *Key World Energy Statistics 2010.* Paris: International Energy Agency, 2010. At http://www.iea.org/textbase/nppdf/free/2010/key_stats_2010.pdf. Accessed April 26, 2011.

International Organization for Migration. "Migration Climate Change and the Environment." n.d. At http://www.iom.int/jahia/Jahia/complex-nexus. Accessed June 30, 2011.

International Scientific Congress on Climate Change. "Climate Change: Global Risks, Challenges, and Decisions Congress: Researchers Present Newest Update on Climate

Change Science." Press release, June 18, 2009. At http://climatecongress.ku.dk/news-room/synthesis_report. Accessed March 27, 2010.

——. "Fighting Global Warming Offers Growth and Development Opportunities." Press release, March 12, 2009. At http://climatecongress.ku.dk/newsroom/mitigation_growth_development. Accessed March 27, 2010.

——. "Rising Sea Levels Set to Have Major Impacts Around the World." Press release, March 10, 2009. At http://climatecongress.ku.dk/newsroom/rising_sealevels. Accessed January 6, 2010.

International Union for Conservation of Nature (IUCN). "Species Extinction: The Facts" (Red List). February 26, 2009. At http://cmsdata.iucn.org/downloads/species_extinction_05_2007.pdf. Accessed July 6, 2011.

——. "Wildlife in a Changing World: An Analysis of the 2008 IUCN Red List of Threatened Species." n.d. At http://data.iucn.org/dbtw-wpd/edocs/RL-2009-001.pdf. Accessed July 6, 2011.

Irwin, Scott. *Is Speculation by Long-Only Index Funds Harmful to Commodity Markets?* Testimony prepared for the U.S. House of Representatives Committee on Agriculture, 108th Cong., 2nd sess., July 20, 2008. At http://www.farmdoc.illinois.edu/irwin/research/House%20Ag%20Testimony,%20July%202008.pdf. Accessed July 1, 2011.

Ismail, A. and B. Manneh. "Benin: Africa Component of STRASA Project Launches Second Phase." *STRASA News* 4, nos. 1–2 (June 2011): 5–6.

Jacobs, Jane. *The Death and Life of Great American Cities.* New York: Random House, 1961.

Jameson, Fredric. *Marxism and Form: Twentieth-Century Dialectical Theories of Literature.* Princeton: Princeton University Press, 1974.

——. *Postmodernism, or, The Cultural Logic of Late Capitalism.* Durham, N.C.: Duke University Press, 1991.

——. *A Singular Modernity: Essay on the Ontology of the Present.* London: Verso, 2002.

——. *Valences of the Dialectic.* London: Verso, 2009.

Jencks, Charles. *The New Paradigm in Architecture: The Language of Post-Modernism.* New Haven: Yale University Press, 2002.

Jenkins, Michael and Ricardo Bayon. "Introduction." In Ricardo Bayon, Amanda Hawn, and Katherine Hamilton, eds., *Voluntary Carbon Markets: An International Business Guide to What They Are and How They Work*, xxi–xxiii. London: Earthscan, 2009.

Jiang, Leiwen and Karen Hardee. "How Do Recent Population Trends Matter in Climate Change?" *Population Action International* 1, no. 1 (April 30, 2009). At http://populationaction.org/wp-content/uploads/2012/01/population_trends_climate_change_FINAL.pdf. Accessed March 20, 2010.

Jones, Jeffrey M. "Americans Prioritize Energy Over the Environment for the First Time." Gallup Poll, April 6, 2010. At http://www.gallup.com/poll/127220/Americans-Prioritize-Energy-Environment-First-Time.aspx. Accessed June 22, 2011.

Kahn, Matthew E. "The Environmental Impact of Suburbanization." *Journal of Policy Analysis and Management* 19, no. 4 (2000): 569–586.

Kaika, Maria. *City of Flows: Modernity, Nature, and the City.* London: Routledge, 2005.

Katz, Peter. *The New Urbanism: Toward an Architecture of Community.* New York: McGraw-Hill, 1994.

Keith, Lierre. *The Vegetarian Myth: Food, Justice, and Sustainability.* Oakland, Calif.: PM Press, 2009.

Kheel, Marti. *Nature Ethics: An Ecofeminist Perspective.* Lanham, Md.: Rowman and Littlefield, 2008.

Klare, Michael T. *Blood and Oil: The Dangers and Consequences of America's Growing Dependency on Imported Oil.* New York: Metropolitan Books, 2004.

Klein, Naomi. *The Shock Doctrine: The Rise of Disaster Capitalism.* New York: Picador, 2008.

Lake, Anthony, Christine Todd Whitman, Princeton N. Lyman, and J. Stephen Morrison. *More Than Humanitarianism: A Strategic U.S. Approach Toward Africa.* New York: Council on Foreign Relations, 2006.

Lang, Chris and Timothy Byakola. *A Funny Place to Store Carbon: UWA–FACE Foundation's Tree Planting Project in Mount Elgon National Park, Uganda.* Montevideo, Uruguay: World Rainforest Movement, December 30, 2006. At http://chrislang. org/2006/12/30/a-funny-place-to-store-carbon-chapter-3. Accessed June 8, 2011.

Langdon, Philip. "A Booming Chicago Readies Itself for Rezoning." *New Urban Network* (March 2003). At http://newurbannetwork.com/article/booming-chicago-readies-itself-rezoning. Accessed April 22, 2011.

Larco, Nico. "Untapped Density: Site Design and the Proliferation of Suburban Multi-family Housing." *Journal of Urbanism: International Research on Placemaking and Urban Sustainability* 2, no. 2 (July 2009): 167–186.

Lees, Loretta. "Gentrification and Social Mixing: Towards an Inclusive Urban Renaissance?" *Urban Studies* 45, no. 12 (November 2008): 2449–2470.

Lemann, Nicholas. *The Promised Land: The Great Black Migration and How It Changed America.* New York: Vintage, 1992.

Levitt, Tom. "Goldman Sachs Makes $1 Billion Profit on Food Price Speculation." *The Ecologist* 40, no. 6 (July 19, 2010). At http://www.theecologist.org/News/news_round_up/542538/goldman_sachs_makes_1_billion_profit_on_food_price_speculation.html. Accessed July 1, 2011.

Lewis, T. R. and J. Cowens. *Cooperation in the Commons: An Application of Repetitious Rivalry.* Vancouver: University of British Columbia, 1983.

Linbaugh, Peter. *The Magna Carta Manifesto: Liberties and Commons for All.* Berkeley: University of California Press, 2009.

Lindstrom, Matthew J. and Hugh Bartling, eds. *Suburban Sprawl: Culture, Theory, and Politics.* Lanham, Md.: Rowman and Littlefield, 2003.

Lipman, Pauline. "The Cultural Politics of Mixed-Income Schools and Housing: A Racialized Discourse of Displacement, Exclusion, and Control." *Anthropology & Education Quarterly* 40, no. 3 (2009): 215–236.

Lipscomb, Hester J., Robin Argue, Mary Anne McDonald, John M. Dement, Carol A. Epling, Tamara James, Steve Wing, and Dana Loomis. "Exploration of Work and Health Disparities Among Black Women Employed in Poultry Processing in the

Rural South." *Environmental Health Perspectives* 113, no. 12 (December 2005). At http://ehp03.nichs.nih.gov/article/fetchArticle.action?articleURI=info%3Adoi%2F1 0.1289%2Fehp.7912. Accessed June 1, 2011.

Lohmann, Larry. "Climate Change Politics After Montreal: Time for a Change." *Foreign Policy in Focus*, Issues: Energy, January 9, 2006. At http://www.fpif.org/articles/climate_politics_after_montreal_time_for_a_change. Accessed June 2, 2011.

Long, Stephen P., Elizabeth A. Ainsworth, Andrew D. B. Leakey, Josef Nösberger, and Donald R. Ort. "Food for Thought: Lower-Than-Expected Crop Yield Stimulation with Rising CO2 Concentrations." *Science* 312, no. 5782 (June 30, 2006): 1918–1921.

Lopez, Russ. "Urban Sprawl and Risk for Being Overweight or Obese." *American Journal of Public Health* 94, no. 9 (September 2004): 1574–1579.

Lovelock, James. *Revenge of Gaia: Earth's Climate Crisis and the Fate of Humanity*. New York: Basic Books, 2006.

——. *The Vanishing Face of Gaia: A Final Warning*. New York: Basic Books, 2009.

Lovins, L. Hunter and Boyd Cohen. *Climate Capitalism: Capitalism in the Age of Climate Change*. New York: Hill and Wang, 2011.

Lukács, Georg. *History and Class Consciousness: Studies in Marxist Dialectics*. Cambridge, Mass.: MIT Press, 1971.

Luttrell, Clifton. *The High Cost of Farm Welfare*. Washington, D.C.: Cato Institute, 1989.

Lynas, Mark. *Six Degrees: Our Future on a Hotter Planet*. Washington, D.C.: National Geographic, 2008.

Maddison, David. "A Cost–Benefit Analysis of Slowing Climate Change." *Energy Policy* 23, nos. 4–5 (1995): 337–346.

Madigan, Charles, ed. *Global Chicago*. Urbana: University of Illinois Press, 2004.

Mahdavy, Hussein. "The Patterns and Problems of Economic Development in Rentier States: The Case of Iran." In M. A. Cook, ed., *Studies in Economic History of the Middle East*, 428–467. London: Oxford University Press, 1970.

Mari Gallagher Research and Consulting Group. *The Chicago Food Desert Progress Report*. Chicago: Mari Gallagher Research and Consulting Group, June 2009. At http://www.marigallagher.com/site_media/dynamic/project_files/ChicagoFoodDesProg2009.pdf. Accessed April 2, 2011.

Martindale, Diane. "Burgers on the Brain." *New Scientist* 177, no. 2380 (February 1, 2003): 26–29.

Martz, John D. *Politics and Petroleum in Ecuador*. New Brunswick, N.J.: Transaction Books, 1987.

Marx, Karl. *Capital*. Vol. 1. Trans. Ben Fowkes. London: Penguin Books, 1990.

——. *The Eighteenth Brumaire of Louis Bonaparte, Surveys from Exile: Political Writings*. Vol. 2. Trans. and ed. David Fernbach. Harmondsworth, U.K.: Penguin, 1973.

——. *Grundrisse*. Trans. Martin Nicolaus. London: Penguin, 1993.

——. *Wage-Labor and Capital*. Trans. Friedrich Engels. New York: International, 1933.

Marx, Karl and Friedrich Engels. *Correspondence 1846–1895*. New York: International, 1935.

Mason, Paul. *Meltdown: The End of the Age of Greed*. London: Verso, 2009.

Massumi, Brian. *Parables for the Virtual: Movement, Affect, Sensation*. Durham, N.C.: Duke University Press, 2002.

May, Todd. *The Political Thought of Jacques Rancière: Creating Equality*. Edinburgh: Edinburgh University Press, 2008.

McBride, Bob. "Broken Heartland: Farm Crisis in the Midwest." *The Nation* 242 (February 8, 1986): 132–133.

McDonough, William and Michael Braungart. *Cradle to Cradle: Remaking the Way We Make Things*. New York: Northpoint Press, 2002.

McGreal, Chris. "George Bush: A Good Man in Africa." *Guardian UK*, February 15, 2008. At http://www.guardian.co.uk/world/2008/feb/15/georgebush.usa. Accessed June 23, 2011.

McKibben, Bill. "Beyond Oil: Activism and Politics." *CounterCurrents*, August 27, 2010. At http://www.countercurrents.org/mckibben270810.htm. Accessed June 22, 2011.

——. *The End of Nature*. New York: Random House, 1989.

——. "Oil Spill Is an Opportunity for Americans." *U.S. News & World Report*, June 28, 2010. Podcast available at http://www.usnews.com/news/best-leaders/articles/2010/06/28/bill-mckibben-oil-spill-is-an-opportunity-for-americans. Accessed June 20, 2011.

McKinney, Merritt. "Flawed Genetic 'Marking' Seen in Cloned Animals." Reuters Health, May 29, 2001.

McLennan, Jason F. *The Philosophy of Sustainable Design*. Kansas City: Ecotone, 2004.

Menin, Sarah and Flora Samuel. *Nature and Space: Aalto and Le Corbusier*. London: Routledge, 2003.

Mercy for Animals. "Maine Egg Farm Investigation." 2008–2009. At http://www.mercyforanimals.org/maine-eggs. Accessed September 30, 2009.

——. "Ohio Dairy Farm Brutality." April–May 2010. At http://www.mercyforanimals.org/ohdairy. Accessed September 30, 2010.

Merill Lynch and Capgemini. *2011 World Wealth Report*. New York: Merill Lynch, 2011. At http://www.ml.com/media/114235.pdf. Accessed July 4, 2011.

Meszaros, Istvan. *The Structural Crisis of Capital*. New York: Monthly Review Press, 2010.

Mintzer, Irving, J. Amber Leonard, and Iván Dario Valencia. *Counting the Gigatonnes: Building Trust in Greenhouse Gas Inventories from the United States and China*. Washington, D.C.: World Wildlife Fund, June 2010, revised September 2010.

Monbiot, George. *Heat: How to Stop the Planet from Burning*. Cambridge, Mass.: South End Press, 2007.

Moore, Soloman. "As Program Moves Poor to Suburbs, Tensions Follow." *New York Times*, August 8, 2008. At http://www.nytimes.com/2008/08/09/us/09housing.html. Accessed March 16, 2011.

Morton, Timothy. *The Ecological Thought*. Cambridge, Mass.: Harvard University Press, 2010.

Mowery, David C., Richard R. Nelson, Bhaven N. Sampat, and Arvids A. Ziedonis. "The Effects of the Bayh–Dole Act on U.S. University Research and Technology Transfer:

An Analysis of Data from Columbia University, the University of California, and Stanford University." Paper presented at the Kennedy School of Government, Harvard University, September 10–12, 1998.

Mufson, Steven. "Iron to Plankton to Carbon Credits." *Washington Post*, July 20, 2007. At http://www.washingtonpost.com/wp-dyn/content/article/2007/07/19/AR200707190 2553.html. Accessed June 4, 2011.

Murtaugh, Paul A. and Michael G. Schlax. "Reproduction and the Carbon Legacies of Individuals." *Global Environmental Change* 19 (2009): 14–20.

Nagel, Joane, Thomas Dietz, and Jeffrey Broadbent. *Workshop on Sociological Perspectives on Climate Change, May 30–31, 2008.* Washington, D.C.: National Science Foundation, 2009. At http://ireswb.cc.ku.edu/~crgc/NSFWorkshop/Readings/NSF_ WkspReport_09.pdf. Accessed July 5, 2011.

National Academy of Sciences, Board of Agriculture and Natural Resources. *Animal Biotechnology: Science Based Concerns.* Washington, D.C.: National Academies Press, 2002.

National Aeronautics and Space Administration (NASA). "Evidence: Climate Change, How Do We Know?" *Climate Change: Vital Signs of the Planet* (a NASA online journal), n.d. At http://climate.nasa.gov/evidence. Accessed July 2, 2011.

National Corn Growers Association. *Understanding the Impact of Higher Corn Prices on Consumer Food Prices.* Chesterfield, Mo.: National Corn Growers Association, March 26, 2007. At http://eerc.ra.utk.edu/etcfc/sefix/dos/FoodCornPrices.pdf. Accessed July 8, 2011.

Negri, Antonio. "Communism: Some Thoughts on the Concept and Practice." In Costas Douzinas and Slavoj Žižek, eds., *The Idea of Communism*, 155–165. London: Verso, 2010.

——. *Time for Revolution.* London: Continuum, 2003.

Negri, Antonio and Félix Guattari. *Communists Like Us: New Spaces of Liberty, New Lines of Alliance.* Trans. Michael Ryan. New York: Semiotext(e), 1990.

Netherlands Environmental Assessment Agency. "China Contributing Two Thirds to Increase in CO2 Emissions." Press release, June 13, 2008. At http://www.pbl.nl/en/ news/pressreleases/2008/20080613ChinacontributingtwothirdstoincreaseinCO2e missions.html. Accessed February 25, 2010.

——. "China Now No.1 in CO2 Emissions; USA in Second Position." Press release, June 19, 2007. At http://www.pbl.nl/en/news/pressreleases/2007/20070619Chinanowno1 inCO2emissionsUSAinsecondposition.html. Accessed February 25, 2009.

Newell, Peter and Matthew Patterson. *Climate Capitalism: Global Warning and the Transformation of the Global Economy.* Cambridge: Cambridge University Press, 2010.

Nordhaus, William. "The Cost of Slowing Climate Change: A Survey." *Energy Journal* 12, no. 1 (1991): 37–65.

Norris, Floyd. "Off the Charts: In '08 Downturn, Some Managed to Eke Out Millions." *New York Times*, July 24, 2010.

"Oil Spill Alters Views on Environmental Protection." Gallup Poll, May 27, 2010. At http://www.gallup.com/poll/137882/oil-spill-alters-views-environmental-protection.aspx. Accessed June 22, 2011.

Olivera, Marcela. "The Cochabamba Water Wars: Marcela Olivera Reflects on the Tenth Anniversary of the Popular Uprising Against Bechtel and the Privatization of the City's Water Supply." Interview by Amy Goodman. *Democracy Now*, April 19, 2010. At http://www.democracynow.org/2010/4/19/the_cochabamba_water_wars_marcella_olivera. Accessed June 5, 2010

Olivera, Oscar, and Tom Lewis, eds. *¡COCHABAMBA!: Water Wars in Bolivia*. Cambridge, Mass.: South End Press, 2004.

Olson, Kevin, ed. *Adding Insult to Injury: Nancy Fraser Debates Her Critics*. London: Verso, 2008.

O'Neill, Brian C., Michael Dalton, Regina Fuchs, Leiwen Jiang, Shonali Pachauri, and Katarina Zigova. "Global Demographic Trends and Future Carbon Emissions." *Proceedings of the National Academy of Sciences* 107, no. 41 (October 12, 2010): 17521–17526.

O'Neill, Brian, F. Landis MacKellar, and Wolfgang Lutz. *Population and Climate Change*. Cambridge: Cambridge University Press, 2001.

Ostrom, Elinor. *Governing the Commons: The Evolution of Institutions for Collective Action*. Cambridge: Cambridge University Press, 1990.

——. "Public Entrepreneurship: A Case Study in Ground Water Basin Management." Ph.D. diss., University of California, Los Angeles, 1965.

Ostrom, Elinor, James Walker, and Roy Gardner. "Covenants with and Without a Sword: Self-Governance Is Possible." *American Political Science Review* 86, no. 2 (June 1992): 404–417.

——. *Rules, Games, and Common-Pool Resources*. Anne Arbor: University of Michigan Press, 1994.

Oswalt, Phillip, ed. *Shrinking Cities*. Vol. 1. Ostfildern-Ruit, Germany: Hatje Cantz, 2005.

Oxfam. "Bold Action Needed Now from G20 Agricultural Ministers to Tackle Causes of Food Price Volatility." Press release, June 21, 2011. At http://www.oxfam.org/en/grow/pressroom/pressrelease/2011-06-21/g20-agricultural-ministers-food-price-volatility. Accessed July 21, 2011.

Pachauri, Rajendra K. Speech at the opening ceremony for the United Nations Framework Convention on Climate Change Intergovernmental Panel on Climate Change, December 1, 2008. At http://www.ipcc.ch/press/index.htm#. Accessed March 4, 2009.

Palaniappan, Meena and Peter H. Gleick. "Peak Water." In Pacific Institute (Peter H. Gleick and others), *The World's Water 2008–2009: The Biennial Report on Freshwater Resources*, 2–16. Washington, D.C.: Island Press, 2008. At http://www.worldwater.org/data20082009/ch01.pdf. Accessed September 1, 2010.

Park, Robert E., Ernest Burgess, and Roderick D. McKenzie. *The City: Suggestions for Investigation of Human Behavior in the Urban Environment*. Chicago: University of Chicago Press, 1967.

Parr, Adrian. *Hijacking Sustainability*. Cambridge, Mass.: MIT Press, 2009.

Parr, Adrian, and Michael Zaretsky, eds. *New Directions in Sustainable Design*. London: Routledge, 2010.

Parra, Francisco. *Oil Politics: A Modern History of Petroleum*. New York: I. B. Taurus, 2004.

Patel, Raj. *Stuffed and Starved: Markets, Power, and the Hidden Battle for the World Food System*. Brooklyn: Melville House, 2007.

——. *The Value of Nothing*. New York: Picador, 2009.

Patel, Sheela, and Jockin Arputham. "An Offer at Partnership or a Promising Conflict in Dharavi, Mumbai?" *Environment and Urbanization* 19, no. 2 (October 2007): 1–8.

Paterson, Matthew. *Understanding Global Environmental Politics: Domination, Accumulation, Resistance*. New York: St. Martin's Press, 2000.

Perkins, John. *Confessions of an Economic Hit Man*. New York: Plume, 2006.

Pew Hispanic Center. *US Population Projections: 2000–2050*. Washington, D.C.: Pew Hispanic Center, February 11, 2008. At http://pewhispanic.org/reports/report.php?ReportID=85. Accessed April 2, 2011.

Pimentel, David and Tad W. Patzek. "Ethanol Production Using Corn, Switchgrass, and Wood; Biodiesel Production Using Soybean and Sunflower." *Natural Resources Research* 14, no. 1 (March 2005): 65–76.

Pollan, Michael. *In Defense of Food: An Eater's Manifesto*. New York: Penguin Books, 2008.

Population Reference Bureau. "2010 World Population Data Sheet." At http://www.prb.org/Publications/Datasheets/2010/2010wpds.aspx. Accessed July 6, 2011.

"ProLinia Announces Collaboration with Smithfield Foods." *PR Newswire*, June 19, 2000. At http://www.thefreelibrary.com/ProLinia+Announces+Collaboration+with+Smithfield+Foods-a062794189. Accessed September 28, 2010.

Public Citizen. *Veolia Environment: A Corporate Profile*. Water for All Campaign. Washington, D.C.: Public Citizen, February 2005. At http://www.citizen.org/documents/Vivendi-USFilter.pdf. Accessed November 1, 2010.

Quandt, Sara A., Joseph G. Grzywacz, Antonio Marin, Lourdes Carrillo, Michael L. Coates, Bless Burke, and Thomas A. Arcury. "Illnesses and Injuries Reported by Latino Poultry Workers in Western North Carolina." *American Journal of Industrial Medicine* 49 (2006): 343–351.

Radeloff, Volker C., Roger B. Hammer, and Susan Stewart. "Rural and Suburban Sprawl in the U.S. Midwest from 1940 to 2000 and Its Relation to Forest Fragmentation." *Conservation Biology* 19, no. 3 (June 2005): 793–805.

Rancière, Jacques. *Disagreement: Politics and Philosophy*. Trans. Julie Rose. Minneapolis: University of Minnesota Press, 1999.

——. *Hatred of Democracy*. Trans. Steve Corcoran. London: Verso, 2006.

——. "The Janus-Face of Politicized Art: Jacques Rancière in Interview with Gabriel Rockhill." In *The Politics of Aesthetics*, trans. Gabriel Rockhill, 49–66. London: Continuum, 2004.

——. *The Politics of Aesthetics*. Trans. Gabriel Rockhill. London: Continuum, 2004

"Recall Expands to More Than Half a Billion Eggs." Associated Press, August 20, 2010. At http://www.msnbc.msn.com/id/38741401. Accessed September 1, 2010.

Regan, Tom. *The Case for Animal Rights*. Berkeley: University of California Press, 1983.

Reidy, Chris. "Colgate Will Buy Tom's of Maine: $100m Deal May Help Boost Sales of Leader in Natural Products Niche." *Boston Globe*, March 22, 2006. At http://www.boston.com/business/articles/2006/03/22/colgate_will_buy_toms_of_maine. Accessed May 1, 2006.

Research Foundation Technology Transfer Office, Colorado State University. "What Is Bayh–Dole and Why Is It Important to Technology Transfer?" October 1999. At http://www.csurf.org/enews/bayhdole_403.html. Accessed October 2, 2010.

Reynolds, Richard. *On Guerrilla Gardening: A Handbook for Gardening Without Boundaries*. London: Bloomsbury, 2008.

Richardson, Amy. "Carbon Credits—Paying to Pollute?" *3rd Degree* 2, no. 7 (October 10, 2006). At http://3degree.cci.ecu.edu.au/articles/view/781. Accessed June 4, 2011.

Richtel, Matt. "Recruiting Plankton to Fight Global Warming." *New York Times*, May 1, 2007. At http://www.nytimes.com/2007/05/01/business/01plankton.html. Accessed June 4, 2011.

Rockstöm, Johan, Will Steffen, Kenvin Noone, Åsa Persson, F. Stuart Chapin III, Eric F. Lambin, Timothy M. Lenton, et al. "Planetary Boundaries: Exploring the Safe Operating Space for Humanity." *Ecology and Society* 14, no. 2, art. 32 (2009). At http://www.stockholmresilience.org/download/18.8615c78125078c8d3380002197/ES-2009-3180.pdf. Accessed July 1, 2011.

——. "A Safe Operating Space for Humanity." *Nature* 461 (September 24, 2009): 472–475.

Rodriguez, C. *Water Management in the Bolivarian Republic of Venezuela*. Washington, D.C.: Embassy of the Bolivarian Republic of Venezuela in the United States, March 2010.

Roig-Franzia, Manuel. "A Culinary and Cultural Staple in Crisis." *Washington Post*, January 27, 2007. At http://www.washingtonpost.com/wp-dyn/content/article/2007/01/26/AR2007012601896.html. Accessed July 7, 2011.

Ross, Michael Lewin. "Does Oil Hinder Democracy?" *World Politics* 53, no. 3 (April 2001): 325–361.

Roy, Arundhati. *The Cost of Living*. New York: Modern Library, 1999.

Runyan, Anne and Marian Marchand, eds. *Gender and Global Restructuring: Sightings, Sites, and Resistances*. London: Routledge, 2010.

Russell, John. "Are Emissions Offsets a Carbon Con?" *Ethical Corporation*, April 1, 2007. At http://www.greenbiz.com/news/reviews_third.cfm?NewsID=34804. Accessed June 1, 2011.

Saad, Lydia. "Americans' Worries About Economy, Budget Top Other Issues." Gallup Poll, March 21, 2011. At http://www.gallup.com/poll/146708/Americans-Worries-Economy-Budget-Top-Issues.aspx. Accessed June 22, 2011.

——. "In U.S., Expanding Energy Output Still Trumps Green Concerns." Gallup Poll, March 16, 2011. At http://www.gallup.com/poll/146651/Expanding-Energy-Output-Trumps-Green-Concerns.aspx. Accessed June 22, 2011.

Sabine, Christopher L., Richard A. Feely, Nicolas Gruber, Robert M. Key, Kitack Lee, John L. Bullister, Rik Wanninkhof, et al. "The Oceanic Sink for Anthropocentric CO2." *Science* 305, no. 5682 (July 2004): 367–371.

Sassen, Saskia. "A Global City." In Charles Madigan, ed., *Global Chicago*, 15–34. Urbana: University of Illinois Press, 2004.

Savitz, Andrew W. *The Triple Bottom Line: How Today's Best-Run Companies Are Achieving Economic, Social, and Environmental Success—and How You Can Too.* San Francisco: Jossy-Bass, 2006.

Schultz, Jim and Melissa Crane Draper. "Conclusion." In Jim Schultz and Melissa Crane Draper, eds., *Dignity and Defiance: Stories from Bolivia's Challenge to Globalization*, 291–296. Berkeley: University of California Press, 2008.

——, eds. *Dignity and Defiance: Stories from Bolivia's Challenge to Globalization.* Berkeley: University of California Press, 2008.

Shanken, Andrew M. *194X: Architecture, Planning, and Consumer Culture on the American Front.* Minneapolis: University of Minnesota Press, 2009.

Sheppard, Kate. "BP's $93 Million Ad Blitz." *Mother Jones*, September 22, 2010. At http://motherjones.com/mojo/2010/09/bps-ad-blitz. Accessed June 1, 2011.

Shiva, Vandana. *Biopiracy: The Plunder of Nature and Knowledge.* Cambridge, Mass.: South End Press, 1997.

Shue, Henry. "After You: May Action by the Rich Be Contingent Upon Action by the Poor?" *Indiana Journal of Global Legal Studies* 1, no. 2 (1994): 343–366.

Simmons, Adele. "Introduction." In Charles Madigan, ed., *Global Chicago*, 7–14. Urbana: University of Illinois Press, 2004.

Simpson, Dick and Tom M. Kelly. "The New Chicago School of Urbanism and the New Daley Machine." *Urban Affairs Review* 44, no. 2 (November 2008): 218–238.

Singer, Peter. *Animal Liberation.* New York: New York Review, 1975.

——. *One World.* New Haven: Yale University Press, 2004.

Sorkin, Michael. "Can New Urbanism Learn from Modernism's Mistakes?" *Metropolis* 18, no. 1 (August–September 1998): 37–39.

Steinfeld, Henning, Pierre Gerber, Tom Wassenaar, Vincent Castel, Mauricio Rosales, and Cees de Haan. *Livestock's Long Shadow: Environmental Issues and Options.* Rome: United Nations Food and Agriculture Organization, 2006.

Stern, Nicholas. *The Economics of Climate Change: The Stern Review.* Cambridge: Cambridge University Press, 2007.

Steve Stice Lab, University of Georgia. "What's Hot in the Stice Lab." n.d. At http://www.biomed.uga.edu/stice. Accessed on October 1, 2010.

Stewart, Kathleen. "Nostalgia—a Polemic." *Cultural Anthropology* 3, no. 3 (August 1988): 227–241.

Story, Louise. "Can Burt's Bees Turn Clorox Green?" *New York Times*, January 6, 2008.

Stull, Donald D. and Michael J. Broadway. *Slaughterhouse Blues: The Meat and Poultry Industry in North America.* Belmont, Calif.: Thomson/Wadsworth, 2003.

Struck, Doug. "Carbon Offsets: How a Vatican Forest Failed to Reduce Global Warming." *Christian Science Monitor*, April 20, 2010. At http://www.csmonitor.com/

Environment/2010/0420/Carbon-offsets-How-a-Vatican-forest-failed-to-reduce-global-warming. Accessed June 4, 2011.

Surin, Kenneth. "The Politics of the Southeast Asian Smog Crisis: A Classic Case of Rentier Capitalism at Work?" In Adrian Parr and Michael Zaretsky, eds., *New Directions in Sustainable Design*, 137–151. London: Routledge, 2010.

SustainLane. "U.S. City Sustainable City Rankings." n.d. At http://www.sustainlane.com/us-city-rankings/overall-rankings. Accessed September 26, 2009.

Switzer, Jason. *Oil and Violence in Sudan*. Winnipeg: International Institute for Sustainable Development and International Union for Conservation of Nature–World Conservation Union Commission on Environmental, Economic, and Social Policy, April 15, 2002.

Swyngedouw, Erik. "Dispossessing H2O: The Contested Terrain of Water Privatization." *Capitalism, Nature, Socialism* 16, no. 1 (March 2005): 81–98.

Torres, Bob. *Making a Killing: The Political Economy of Animal Rights*. Oakland, Calif.: AK Press, 2007.

Tracey, David. *Guerrilla Gardening: A Manualfesto*. Gabriola Island, Canada: New Society, 2007.

"Transcript: Vice Presidential Debate." *New York Times*, October 2, 2008. At http://elections.nytimes.com/2008/president/debates/transcripts/vice-presidential-debate.html. Accessed June 23, 2011.

United Nations. *Kyoto Protocol to the United Nations Framework Convention on Climate Change*. 1998. At http://unfccc.int/resource/docs/convkp/kpeng.pdf. Accessed May 30, 2011.

——. "World Population to Exceed 9 Billion by 2050." Press release, March 11, 2009. At http://www.un.org/esa/population/publications/wpp2008/pressrelease.pdf. Accessed March 20, 2010.

United Nations Department of Economic and Social Affairs, Population Division. *World Population to 2300*. New York: United Nations Department of Economic and Social Affairs, 2004. At http://www.un.org/esa/population/publications/longrange2/WorldPop2300final.pdf. Accessed May 4, 2010.

United Nations Environment Programme (UNEP). *Buildings and Climate Change: Status, Challenges, and Opportunities*. Malta: UNEP, 2007.

——. *Global Environmental Outlook 4: Environment for Development*. Malta: UNEP, 2007.

——. "World Food Supply: Food from Animal Feed." n.d. At http://www.grida.no/publications/rr/food-crisis/page/3565.aspx. Accessed July 10, 2011.

United Nations Food and Agriculture Organization (UN FAO). *Climate Change and Food Security: A Framework Document*. Rome: UN FAO, 2008.

——. *The State of Food Insecurity in the World: Addressing Food Insecurity in Protracted Crisis*. Rome: UN FAO, 2010. At http://www.fao.org/docrep/013/i1683e/i1683e.pdf. Accessed July 1, 2011.

United Nations Framework Convention on Climate Change. "Clean Development

Mechanism." n.d. At http://unfccc.int/kyoto_protocol/mechanisms/clean_develop ment_mechanism/items/2718.php. Accessed May 30, 2011.

———. "CDM-Certified Projects." n.d. At http://cdm.unfccc.int/Projects/registered.html. Accessed May 30, 2011.

———. "Joint Implementation." n.d. At http://unfccc.int/kyoto_protocol/mechanisms/joint_implementation/items/1674.php. Accessed May 30, 2011.

United Nations High Commissioner for Refugees (UNHCR) and U.S. Committee for Refugees and Immigrants. *U.S. Committee Mid Year Country Report: Sudan.* Geneva: UNHCR, October 2, 2001. At http://www.unhcr.org/refworld/country,,USCRI,,SDN ,456d621e2,3c56c1161c,0.html. Accessed June 23, 2011.

United Nations Human Settlement Programme (UN-HABITAT). *Hot Cities: Battle-ground for Climate Change.* Nairobi: UN-HABITAT, March 2011. At http://www .unhabitat.org/downloads/docs/GRHS2011/P1HotCities.pdf. Accessed April 2, 2011.

United Nations Population Fund. *The State of World Population 2009.* New York: United Nations Population Fund, 2009. At http://www.unfpa.org/swp/2009. Accessed March 31, 2010.

United Nations Statistics Division. *Environmental Indicators: Greenhouse Gas Emissions 2007.* New York: United Nations Statistics Division, 2007. At http://unstats.un.org/unsd/environment/air_co2_emissions.htm. Accessed April 26, 2011.

United Nations Water. "Statistics, Graphs, and Maps." n.d. At http://www.unwater.org/statistics.html. Accessed on September 1, 2010.

———. "Water Use." n.d. At http://www.unwater.org/statistics_use.html. Accessed May 5, 2010.

U.S. Agency for International Development (USAID). "USAID Responds to Global Food Crisis." May 22, 2009. At http://www.usaid.gov/our_work/humanitarian_ assistance/foodcrisis. Accessed July 10, 2011.

U.S. Bureau of Labor Statistics. *Local Area Unemployment Statistics: Unemployment Rates for States.* Washington, D.C.: U.S. Bureau of Labor Statistics, n.d. At http://www.bls.gov/web/laumstrk.htm. Accessed September 23, 2009.

U.S. Census Bureau. *American Community Survey.* Washington, D.C.: U.S. Census Bureau, 2009. At http://www.census.gov/acs/www. Accessed September 1, 2010.

U.S. Department of Agriculture and National Agriculture Statistics Services. "Milk Production." September 17, 2010. At http://usda.mannlib.cornell.edu/usda/current/MilkProd/MilkProd-09-17-2010.pdf. Accessed September 25, 2010.

U.S. Department of Housing and Urban Development. "About HOPE VI." 2009. At http://www.hud.gov/offices/pih/programs/ph/hope6/about. Accessed April 26, 2011.

U.S. Energy Information Administration. *Oil Consumption.* Washington, D.C.: U.S. Energy Information Administration, 2009. At http://www.eia.gov/countries/index.cfm?view=consumption. Accessed June 23, 2011.

———. *Oil: Crude and Petroleum Products Explained.* Washington, D.C.: U.S. Energy Information Administration, 2010. At http://www.eia.gov/energyexplained/index.cfm?page=oil_home#tab2. Accessed June 22, 2011.

U.S. Environmental Protection Agency. "Glossary of Climate Change Terms." n.d. At http://www.epa.gov/climatechange/glossary.html. Accessed June 3, 2011.

U.S. Food and Drug Administration. "Animal Cloning." April 26, 2010. At http://www.fda .gov/AnimalVeterinary/SafetyHealth/AnimalCloning/default.htm. Accessed October 1, 2010.

U.S. Government Accountability Office (GAO). *School Lunch Program: Efforts Needed to Improve Nutrition and Encourage Healthy Eating.* Washington, D.C.: U.S. GAO, May 2003.

——. *Workplace Safety in the Meat and Poultry Industry, While Improving, Could Be Further Strengthened.* A report to the Ranking Minority Member, Committee on Health, Education, Labor, and Pensions, U.S. Senate. Washington, D.C.: U.S. GAO, January 2005.

U.S. Green Building Council. "Buildings and Climate Change." n.d. At http://www .usgbc.org/DisplayPage.aspx?CMSPageID=2124. Accessed April 1, 2011.

U.S. National Oceanic and Atmospheric Administration. Earth System Research Laboratory data. At ftp://ftp.cmdl.noaa.gov/ccg/co2/trends/co2_annmean_mlo.txt. Accessed May 30, 2011.

Van Ingen, T. and C. Kawau. *Involvement of Women in Planning and Management in Tanga Region, Tanzania.* Gland, Switzerland: International Union for Conservation of Nature and World Conservation Union, 2003.

ViaGen. "ViaGen Acquires Livestock Pioneer ProLinia." Press release, June 30, 2003. At http://www.viagen.com/news/viagen-acquires-livestock-pioneer-prolinia. Accessed June 1, 2011.

Vicziany, Marika. "Coercion in a Soft State: The Family Planning Program of India, Part One: The Myth of Voluntarism." *Pacific Affairs* 55, no. 3 (1982): 373–402.

Walters, Billy. Interviewed by Lara Logan. *60 Minutes,* CBS, January 16, 2011.

Wara, Michael. "Is the Global Carbon Market Working?" *Nature* 445 (February 8, 2007): 595–596.

Warwick, David. *Bitter Pills: Population Policies and Their Implementation in Eight Developing Countries.* Cambridge: Cambridge University Press, 1982.

Water Footprint Network. Home page. At http://www.waterfootprint.org/?page=files/ home. Accessed May 5, 2010.

Watkins, Kevin. *Summary Human Development Report 2005.* New York: United Nations Development Program, 2005.

Watts, Michael. *Imperial Oil: The Anatomy of a Nigerian Oil Insurgency.* Economies of Violence Working Papers, Working Paper no. 17. Berkeley: Institute of International Studies, University of California, 2008.

Welch, Jarrod R., Jeffrey R. Vincent, Maximilian Auffhammer, Piedad F. Moya, Achim Dobermann, and David Dawe. "Rice Yields in Tropical/Subtropical Asia Exhibit Large but Opposing Sensitivities to Minimum and Maximum Temperatures." *Proceedings of the National Academy of Sciences* 107, no. 33 (August 17, 2010): 14562–14567.

Wesley, E. and F. Peterson. "The Ethics of Burden Sharing in the Global Greenhouse." *Journal of Agricultural and Environmental Ethics* 11 (1999): 167–196.

Willadsen, S. M., R. E. Janzen, R. J. McAlister, B. F. Shea, G. Hamilton, and D. McDermand. "The Viability of Late Morulae and Blastocysts Produced by Nuclear Transplantation in Cattle." *Theriogenology* 35, no. 1 (January 1991): 161–170.

Wilson, William Julius. *When Work Disappears: The World of the New Urban Poor.* New York: Vintage Books, 1997.

Wolff, Edward N. *Recent Trends in Household Wealth in the United States: Rising Debt and the Middle-Class Squeeze—an Update to 2007.* Working Paper no. 589. Annandale-on-Hudson, N.Y.: Levy Economics Institute, Bard College, March 2010. At http://www.levyinstitute.org/pubs/wp_589.pdf. Accessed September 12, 2010.

——. *Top Heavy: A Study of Increasing Inequality of Wealth in America.* New York: Twentieth Century Fund Press, 1995.

——. "The Wealth Divide: The Growing Gap in the United States Between the Rich and the Rest." *Multinational Monitor* 24, no. 5 (May 2003). At http://multinationalmonitor.org/mm2003/03may/may03interviewswolff.html. Accessed February 18, 2012.

World Bank. "How We Classify Countries." n.d. At http://data.worldbank.org/about/country-classifications. Accessed January 11, 2011.

——. "Outlook Is for Steady but Slower Growth in 2011 and 2012." In *Global Economic Prospects 2011*, a report. Washington, D.C.: World Bank, January 12, 2011. At http://go.worldbank.org/5AYIR3UW70. Accessed June 1, 2011.

——. "Retracting Glacier Impacts Economic Outlook in Tropical Andes." April 23, 2008. At http://go.worldbank.org/W5C3YWZFG0. Accessed June 5, 2010.

——. *Water Resources Management: A World Bank Policy Paper.* Washington, D.C.: World Bank, 1993.

World Commission on Environment and Development. *Our Common Future.* Oxford: Oxford University Press, 1990.

World Economic Forum Water Initiative. "Managing Our Future Water Needs for Agriculture, Industry, Human Health, and the Environment." Draft for discussion at the World Economic Forum Annual Meeting, January 2009. At http://www3.weforum.org/docs/WEF_ManagingFutureWater%20Needs_DiscussionDocument_2008.pdf. Accessed January 15, 2011.

World Health Organization (WHO). "Human Rights–Based Approach to Health." n.d. At http://www.who.int/trade/glossary/story054/en/index.html, under "Trade, Foreign Policy, Diplomacy, and Health." Accessed July 8, 2011.

——. *Protecting Health from Climate Change: Connecting Science, Policy, and People.* Geneva: WHO, 2009. At http://whqlibdoc.who.int/publications/2009/9789241598880_eng.pdf. Accessed July 6, 2011.

World Resources Institute. "Water: Critical Shortages Ahead?" n.d. At http://www.wri.org/publication/content/8261. Accessed June 15, 2010.

Yates, Douglas A. *The Rentier State in Africa: Oil Rent Dependency and Neocolonialism in the Republic of Gabon.* Trenton, N.J.: Africa World Press, 1996.

Yonek, Juliet and Romana Hasnain-Wynia. *A Profile of Health and Health Resources Within Chicago's 77 Community Areas.* Chicago: Northwestern University Feinberg School of Medicine, Center for Healthcare Equity/Institute for Healthcare

Studies, 2011. At http://chicagohealth77.org/uploads/Chicago-Health-Resources-Report-2011-0811.pdf. Accessed February 20, 2012.

Zahniser, Steven and William Coyle. *U.S.–Mexico Corn Trade During the NAFTA Era: New Twists to an Old Story*. FDS-04D-01. Washington, D.C.: U.S. Department of Agriculture, May 2004. At http://ip.cals.cornell.edu/courses/iard602/2007spring/mexico/mexico/USMEX_Corn_Trade.pdf. Accessed July 8, 2011.

Zalasiewicz, Jan, Mark Williams, Will Steffen, and Paul Crutzen. "The New World of the Anthropocene." *Environmental Science and Technology* 44, no. 7 (2010): 2228–2231.

Zaretsky, Michael. "LEED After Ten Years." In Adrian Parr and Michael Zaretsky, eds., *New Directions in Sustainable Design*, 185–190. London: Routledge, 2010.

Žižek, Slavoj. *Living in the End Times*. London: Verso, 2010.

——. *Violence*. New York: Picador, 2008.

INDEX

"ability-to-pay principle," 9

accumulation through dispossession, 112

acid rain, 24, 29, 154n18

Adams, Carol, 92, 94, 95

"additionality" requirement (Kyoto Protocol), 26, 30

adivasi, 31, 33, 37, 38

Advanced Cell Technology, 106

aesthetics, 138, 177n17

affermage, 59

Africa: food production in, 75, 76; rice production in, 79; solar power in, 16; U.S. HIV/AIDS programs for, 16, 141

African American communities: food deserts, 126; public housing in Chicago, 119–120, 121, 122

African Americans, income disparities of, 123

agriculture: climate change and, 75; conservation agriculture, 78–79, 87; genetic engineering, 79, 80; international trade, 102

alterity, 177n17

Anglian Water, 59

animal breeding-management programs, 104–105

animal cruelty, 90–91, 98, 109

animal rights, Marxist theory and, 95

animals: animal abuse, 90–91; breeding-management programs, 104–105; cloning, 105–106, 171nn49, 53; commodification of, 95; empathy and care, 94; environmental change and, 88–110; feminist-animal liberation, 92, 94, 171n48; meatpacking, 98–100, 170nn32, 34, 35, 173n69; Mercy for Animals undercover videos, 90; moral value of, 93; objectification of, 94; reproduction, 104–106; slaughter process, 109, 169n28; speciesism, 93, 94; species biodiversity loss, 9, 13, 53, 156n2, 159n3

Annex 1/2 countries, 25

Anthropocene age, 3, 149n6

Argentina, conservation agriculture, 78

AricaRice (program), 79

Arrighi, Giovanni, 159n32

artificial insemination, of farm animals, 104–105

Asia, climate change and agriculture, 75, 76

Attfield, Robin, 10, 17

Bangladesh, floods in, 52

Banzer, Hugo, 57

Barlow, Maude, 59